Hermann Rump

Laboratory Manual
for the Examination
of Water, Waste Water
and Soil

WILEY-VCH

The Reference Work for Seawater Analysis:
K. Grasshoff, K. Kremling, M. Ehrhardt (Eds.)
Methods of Seawater Analysis
1999. ISBN 3-527-29589-5

Hans Hermann Rump

Laboratory Manual for the Examination of Water, Waste Water and Soil

Third, completely revised edition

English Translation by Elisabeth J. Grayson

 **WILEY-VCH**

Weinheim · New York · Chichester · Brisbane · Singapore · Toronto

Prof. Dr. Hans Hermann Rump
Kreditanstalt für Wiederaufbau
Hauptabteilung Technik
Palmengartenstr. 5–9
D-60325 Frankfurt

Library of Congress Card No.: applied for

British Library Cataloguing-in-Publication Data: A CIP catalogue record of this book is available from the British Library

Deutsche Bibliothek Cataloguing-in-Publication Data: Ein Titeldatensatz für diese Publikation ist bei der Deutschen Bibliothek erhältlich

© WILEY-VCH Verlag GmbH. D-69469 Weinheim (Federal Republic of Germany), 1999

Printed on acid-free and chlorine-free paper.

Composition: M-O-P-S Multi Original Print Service, D-50259 Pulheim-Brauweiler
Printing: Strauss Offsetdruck, D-69509 Mörlenbach
Bookbinding: Großbuchbinderei J. Schäffer, D-67269 Grünstadt

Printed in the Federal Republic of Germany.

Preface to the Third Edition

The laboratory handbook was conceived 15 years ago on the initiative of the Deutsche Gesellschaft für Technische Zusammenarbeit (German Agency for Technical Co-operation) (GTZ) to provide reliable laboratory methods suitable for teaching and plant laboratories. At that time procedures involving the use of sophisticated apparatus were deliberately not included. Soil science techniques were only included where they were related to the protection of bodies of water and agricultural irrigation. This basic principle has essentially been followed in the third edition, although, because many laboratories are now better equipped, some more complex methods have been included.

All parts of the text have been rewritten and the data brought up to date. However, the layout of the book has remained the same. The description of the work which needs to be carried out prior to the analysis has been extended and improved, as comments on the last edition indicated that a detailed description of sampling methods and evaluation of test results is also expected as well as that of purely analytical procedures. Incorporation of an area of work into the overall analytical quality assurance procedures can have a positive effect on the results. However, errors in the planning of work and in sampling can still not be rectified even if the analysis procedures are perfect. The chapter on *Safety in the Laboratory* has again been put first to emphasise the importance of protection of the workplace and the environment. New sections of this chapter are the treatment and disposal of waste water, waste air and laboratory waste, eco-audits for laboratories and industrial medicine aspects. The chapter on *Quality Control* is new. The chapters on *Requirements for Analytical Methods and Organisation of Sample Collection Programmes and Sampling Techniques* have basically been rewritten. *Field Measurements and Laboratory Measurements* are presented in the same form as before but a large number of important methods, some of them new, have been added. The chapter entitled *Interpretation of Test Results* lists threshold and recommended values which are valid at the time of writing. The list of *References* allows particular areas to be consulted in greater depth and summarises the most important legal and technical regulations. The *Index* has also been extended to facilitate looking up a particular subject.

The manual is intended mainly for specialists working in the area of water chemistry, water and waste water technology and environmental protection. It was pleasing that the previous editions also proved useful to teachers and students in scientific and technical departments of universities and colleges.

There has been much stimulating input and many suggestions for improvements for this edition. In particular I would like to thank Drs V Neitzel and B Scholz and my former colleagues at the Fresenius Institute, who have given me important information on quality assurance and new analytical methods. I thank my wife Lisa and my son Sebastian for helping to prepare the manuscript. My thanks also go to Wiley-VCH for their constructive and trusting collaboration and to the GTZ for supporting the work for so long.

Frankfurt, May 1999   Hans Hermann Rump

Table of contents

List of abbreviations

List of Abbreviations

AAS	atomic absorption spectrometry
AOX	adsorbable organic halogen compounds
ASTM	American Society for Testing and Materials
ATV	Abwassertechnische Vereinigung (German Wastewater Association)
BOD_5	biochemical oxygen demand in 5 days
cm	centimetre
COD	chemical oxygen demand
conc.	concentrated
DB	4-(2,4-dichlorophenoxy)butanoic acid
DDT	1,1'-(2,2,2-trichloroethylidene)bis[4-chlorobenzene]
DIN	Deutsche Industrienorm (German standard)
DOC	dissolved organic carbon
DVGW	Deutscher Verein des Gas- und Wasserfachs
DVWK	Deutscher Verband für Wasserwirtschaft und Kulturbau
EC	European Community
EDTA	ethylenediaminetetraacetic acid
EEC	European Economic Community
EN	Europäische Norm (European standard)
FAO	Food and Agricultural Organisation
FNU	formazine nephelometric units
FTU	formazine turbidity units
g	gramme
GLP	Good Laboratory Practice
h	hour
hPa	hectopascal
IR	infrared
ISO	International Standards Organisation
kg	kilogramme
l	litre
LAWA	Länderarbeitsgemeinschaft Wasser (a working group for water set up by the German regional governments)
LC	lethal concentration
LD_{50}	lethal dose for 50% of the organisms tested
m	metre
MAK	maximale Arbeitsplatzkonzentration (maximum workplace concentration)
MBAS	methylene blue active substance
MCPA	(4-chloro-2-methylphenoxy)acetic acid
min	minute
µg	microgramme
ml	millilitre

µm	micrometre
mV	millivolt
nm	nanometre
NTU	nephelometric turbidity units
PCB	polychlorinated biphenyl
S	Siemens ($= \Omega^{-1}$)
SAC	spectroscopic absorption coefficient
SAR	sodium adsorption ratio
sec	second
TIC	total inorganic carbon
TRGS	Technische Regeln für Gefahrstoffe (Technical Rules on Hazardous Materials)
TTC	triphenyltetrazolium chloride
TVO	Trinkwasserverordnung (German ordinance on drinking water)
UV	ultraviolet
VBG	Verband der Berufsgenossenschaften or Vorschrift der Berufsgenossenschaft (Employment Accident Insurance Fund)
WHG	Wasserhaushaltgesetz (Federal Water Act)
WHO	World Health Organisation
Ω	Ohm

1 Safety in the Laboratory

A sound understanding of laboratory equipment and test methods and competence in measures for the protection of health and the workplace, and for protection against fire can significantly reduce the danger of accidents occurring in chemical laboratories. The following advice should make responsible work in the laboratory easier and should therefore be studied carefully, particularly by inexperienced personnel.

Typical hazards in the chemical laboratory lie in the use of

– substances which are flammable, explosive or harmful to health,
– high pressures and temperatures,
– electricity.

Specific accidents in the laboratory result from:
poisoning; fires and explosions through handling flammable gases, vapours, or solid materials; burns from hot substances or liquids and corrosive burns from contact with acids and alkalis; explosion of pressurised containers; the effects of electric currents on the human body.

The golden rule is:

Laboratory work must always be carried out with the greatest care and attention.

1.1 Basic Rules for Laboratory Safety

Work with laboratory chemicals and equipment becomes less dangerous if the following rules are adhered to:

– Only qualified personnel should be asked to handle dangerous reagents and equipment.
– Safety goggles must always be worn and, when necessary, protective gloves and other protective clothing.
– All work should be carried out in well ventilated areas; an efficient fume hood must be employed when dangerous gases or fumes are released.
– Chemicals should not be permitted to come into contact with the eyes, mucous membranes or skin.
– If chemicals do come into contact with the eyes, they should be rinsed thoroughly with water or, preferably, with a special solution, with the patient in a reclining position.
– Splashes of dangerous liquids on the skin should first be removed with a dry cloth, and the affected area then rinsed for a prolonged period with cold water; it should then be washed with warm water and soap.
– Clothing contaminated with corrosive substances should be removed immediately.

1.2   Handling of Chemicals and Samples

Many chemicals and samples can present a hazard for both laboratory personnel and the environment. They can be toxic, harmful to health, corrosive, irritant, flammable, highly flammable, or pathogenic. The substitution of hazardous chemicals by less or non-hazardous substances is the safest way of eliminating the danger posed by them. If this is not possible, substances harmful to health should be used in closed apparatus where feasible; otherwise fume hoods should be used (see Section 1.8).

Chemicals must be stored in suitable containers and these must be labelled showing the contents, hazard class and hazard symbols. Storage of unnecessarily large quantities of chemicals should be avoided.

In the analytical procedures described in this book, particular hazards are indicated by symbols at the appropriate points (Fig. 1).

Usage

Dangerous chemicals and solvents may not be handled in breakable containers exceeding 5 l capacity. Exceptions are only permitted if special protective measures are taken, such as the use of trapping containers together with absorptive and fire extinguishing materials. In processing samples containing pathogenic bacteria (waste water, sewage sludge, or incubated culture media) special protective measures are necessary (use of digesters, gloves and mouth protection). Immunisation against tetanus and hepatitis is advisable.

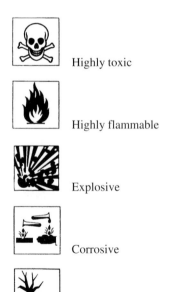

Highly toxic

Highly flammable

Explosive

Corrosive

Hazardous to the environment

Fig. 1: Hazard symbols according to the Hazardous Materials Ordinance

Transport

Breakable containers should never be carried by the neck only, but should also be supported from below. For transport over longer distances, secure packing and carriage should be ensured, for example in buckets or boxes. Containers removed from cold rooms are usually very slippery to handle because of the condensation of water.

Storage

In general, laboratory chemicals and samples should be stored under cool and dry conditions. Larger quantities of dangerous chemicals should never be kept in the laboratory itself, but in a chemical store constructed according to official regulations. Chemical stores must be constructed in bunded areas with a threshold and walls at least 10 cm high, and must not drain into the sewage system. Emergency showers, sprinkler systems, fire extinguishers and personal protective equipment, such as gloves, safety shoes and chemical-resistant clothing, must be available. When equipping chemical and sample stores, the relevant regulations must be observed.

It is essential that explosive and flammable substances are stored separately from one another. Cold storage is recommended for the following:

– flammable gases in pressurised containers,
– liquids whose boiling points can easily be reached under unsuitable storage conditions, e.g. acetone, diethyl ether, pentane, hexane, petroleum ether, carbon disulfide and dichloromethane.

The following substances with particular hazard symbols on the container should be stored in a lockable room or cupboard if possible:

– cyanides,
– mercury and its compounds,
– arsenic and its compounds,
– alkali metals,
– thallium compounds,
– uranium compounds,
– phosphorus,
– pesticides.

Chemicals which emit corrosive fumes must be stored in well ventilated areas, e.g.:

– hydrofluoric acid,
– nitric acid,
– hydrochloric acid,
– aqueous ammonia,
– bromine.

The following substances must be stored separately from certain other substances (Table 1).

Table 1: Chemical substances to be isolated from others

Substance	To be kept separate from
Activated charcoal	Oxidising agents, calcium hypochlorite
Alkali metals	Water, carbon dioxide, chlorinated hydrocarbons, halogens
Ammonia gas	Mercury, halogens
Ammonium nitrate	Acids, metal powders, flammable liquids, sulfur, finely divided organic materials
Flammable liquids	Oxidising agents such as ammonium nitrate, chromium(VI) oxide, nitric acid, sodium peroxide, halogens
Chlorates	Ammonium salts, acids, metal powders, sulfur, finely divided organic materials
Chromium(VI) oxide	Flammable liquids
Cyanides	Acids
Hydrofluoric acid	Ammonia
Potassium permanganate	Glycerol, ethylene glycol, conc. sulfuric acid
Copper	Acetylene, hydrogen peroxide
Sodium peroxide	Flammable liquids
Oxalic acid	Silver, mercury
Perchloric acid	Alcohols, paper, wood, acetic anhydride, bismuth and its alloys
Phosphorus	Sulfur, chlorates
Mercury	Acetylene, ammonia
Nitric acid (conc.)	Acetic acid, chromium(VI) oxide, hydrogen sulfide, flammable liquids and gases
Sulfuric acid (conc.)	Potassium chlorate, potassium perchlorate, potassium permanganate
Silver	Acetylene, oxalic acid, tartaric acid, ammonium compounds
Hydrogen peroxide	Metals and metal salts, organic compounds

1.3   Fire Hazards and Fire Prevention

Fires require all the following conditions:

–   combustible material (solids, gases, vapours),
–   oxygen, air, or other oxidants,
–   heat (flame, spark, heating element, hotplate).

For liquids to ignite, the temperature must be above the flashpoint and the ignition temperature must be reached rapidly, so that the vapour produced can react with oxygen. The flashpoint and the ignition temperature are specific for a particular substance. The flashpoint of a flammable liquid is the lowest temperature at which the liquid forms a vapour/air mixture under defined conditions which can be ignited by an external source. The ignition temperature is the lowest temperature under defined experimental conditions at which the material spontaneously ignites at normal pressure. The values for flashpoints and ignition temperatures can be obtained from reference books (see References).

 – Liquids with a *flashpoint < 21 °C* are classed as highly flammable and include:
   immiscible with water:
   petrol, benzene, diethyl ether, carbon disulfide, ethyl acetate, toluene.
   miscible with water:
   methanol, ethanol, propanol, isopropanol, pyridine, acetone, tetrahydrofuran.
 – Liquids with a *flashpoint between 21 and 55 °C:*
   butanol, butyl acetate, chlorobenzene, amyl alcohol, acetic anhydride, xylenes.
 – Liquids with a *flashpoint between 55 and 100 °C:*
   dichlorobenzene, cresols, heating oil, nitrobenzene, phenol, paraffin oil.

   For storage of these liquids see Section 1.2.

   Mixtures of flammable materials with air cannot be ignited in all proportions. Ignitable mixtures are characterised by their upper and lower explosion limits, given as the volume fraction (%) or in g/m^3 (see References).

Rules for Fire Prevention
 – Exercise the greatest care on heating more than 50 ml of flammable liquids.
 – It is essential to avoid the spreading of flammable vapours. Vapours are often heavier than air and can spread over several metres. Ignition sources (bunsen burners and mantles) should be removed, even when concealed or at some distance from the vapours.
 – Electrostatic charges can start fires through sparking. Charging can occur on filling glass or plastic containers with non-conducting liquids (e.g. acetone, ether, carbon disulfide, toluene). Liquids should therefore be poured slowly avoiding free fall and, if possible, using a funnel reaching the container bottom. Two conducting containers should be earthed to each other, as should two non-conducting containers.
 – No highly flammable liquid should be stored in a refrigerator, as ignition of vapours by sparks (light or thermostat) is possible.
 – Bumping should be avoided during distillation by using antibumping granules at normal pressure and capillaries in the case of vacuum distillation.

Protective Equipment and Fire Extinguishers
 – Escape routes must be provided, clearly marked and kept free from hindrances; a second exit from the laboratory must be available.
 – Fire alarms must be installed and the emergency number marked near the telephone.
 – At least one safety shower per laboratory should be provided.
 – Fireproof blankets must be readily available.

– Sand buckets must be available.
– Portable carbon dioxide extinguishing equipment and hand fire extinguishers (CO_2 or powder) must be within easy reach.
– Regular fire practice must be carried out.
– All equipment and materials for fire fighting and protection, as well as fire alarms should be marked in red.

Fire Fighting Procedure
– Remove injured persons from the danger zone.
– Persons with burning clothing should be rapidly wrapped in an extinguishing blanket on the floor or sprayed with a carbon dioxide extinguisher (not in the face!) or pulled under the laboratory shower.
– The fire alarm should be raised.
– Flammable material and gas cylinders should be removed.
– The gas supply should be switched off at the mains.
– In case of fire involving electrical equipment the electricity supply should be switched off before fire fighting begins.
– The fire should be tackled, or, if this is unsuccessful, the building should be evacuated.

Advice on Fire Fighting in the Laboratory
For most laboratory fires carbon dioxide extinguishers are sufficient. They leave no residues and therefore cause no contamination of rooms or damage to sensitive apparatus. In addition carbon dioxide is chemically relatively unreactive and can also be used with electrical equipment. After the fire has been extinguished the room should be thoroughly ventilated immediately to avoid the danger of suffocation. Fires involving alkali metals or lithium aluminium hydride should not be fought with water under any circumstances, but with sand or cement powder. For flammable liquids carbon dioxide or powder extinguishers should be used. The use of halon extinguishers is no longer permitted.

1.4   Electricity

If an electric current flows through the human body through contact with current-carrying equipment, severe burns can result. In addition, arrhythmia or even heart failure can occur. The current which flows through the body determines the extent of injury: the greater the contact potential and the lower the resistance at the point of contact, the higher the current. With wet hands and a conducting floor a potential as small as 50 V can be dangerous, as large currents are produced compared with the safer conditions of dryness and insulation. The duration of current flow is also important. For example, the average resistance of the human body is $1\,300\,\Omega$. With a good contact on the hands or feet, a current of ca. 170 mA flows through the body at a potential of 220 V. With an alternating current death can occur within a few seconds. In the case of direct current particular care is recommended because even lower voltages can be fatal due to the resulting electrolytic processes in the body.

Advice on the Use of Electricity
- All equipment with more than 50 V potential is classified as high tension equipment. Installation and repairs must be carried out by properly qualified specialists according to the VDE regulations.
- Sockets, plugs, cables and equipment should have their insulation tested before use. Defective components and all wet equipment pose dangers on contact with the body.
- It should be possible to shut of the electricity supply in each room via a readily accessible master switch.
- All apparatus should have earthed contact fuses. These must be tested periodically.

1.5   First Aid

Here only the most important advice can be given. First aid by amateurs is no substitute for proper medical treatment but is only an emergency measure. A practical first aid course should be obligatory for all laboratory personnel.

Equipment
- addresses and telephone numbers of emergency doctors, hospitals and burns clinics,
- wall chart with first aid instructions,
- first aid kit,
- separate eye-wash bottles for damage to the eyes by acids and alkalis.

Procedure

   Corrosive Burns
- Skin burns: the affected areas are washed thoroughly with water.
- Bromine burns: wash with paraffin oil or ethanol.
- Hydrofluoric acid burns: wash with 2% ammonia solution or dilute sodium bicarbonate.
- Iodine burns: wash with 1% sodium thiosulfate solution.
- Eye burns: hold the affected eye wide open with both hands and wash with water for about 10 min. The eye should be rolled in all directions during washing.

   Cuts
- Do not touch or wash the wound but cover with a sterile dressing (plaster, bandage).

   Burns
- These must be kept sterile and covered with a special burn dressing (not for facial areas). Burns caused by phosphorus should be bathed with sodium bicarbonate solution.

   Poisoning
- If the patient is conscious, vomiting should be induced. Activated charcoal tablets may also be given. It is essential to retain any remaining poison and vomited material. Lie the patient on his/her side.

   Action on Loss of Consciousness
- Lie the patient on his/her side.
- Bend the head backwards with the face towards the floor.
- Monitor pulse and breathing.

– If breathing stops give artificial respiration: first blow air strongly 20 times into the mouth or nose, then wait for ca. 30 sec, and then continue to blow in air at the normal rate. Experienced first-aiders can carry out heart massage. These measures should be continued until the doctor arrives.

1.6   Disposal of Dangerous Laboratory Waste

Quantities of waste from laboratories are relatively small overall, but their size and composition can vary. These or similar regulations apply to the disposal of waste from industrial, teaching and research laboratories:

– Closed Substance Cycle and Waste Management Act,
– Ordinance on the Classification of Waste Requiring Special Supervision,
– Ordinance on the Furnishing of Proof,
– Employment Accident Insurance Fund, chemical laboratory guidelines
– Technical Rules on Hazardous Materials (TRGS), 201: labelling of waste materials.

The assignment of laboratory waste to plants within the disposal site is carried out according to the Technical Instructions on Waste. Proof of disposal is sent to the authority responsible after processing by the disposal personnel and the drawing up of the declaration of acceptance, as in the case of waste from external sources.

For some types of waste a single classification according to the laws governing waste and hazardous substances is not sufficient. Here repeated random sampling is necessary to test whether the waste still satisfies the criteria of the authorised proof of disposal and the classification with regard to its nature and composition, according to the laws governing hazardous materials.

In exceptional cases, if waste cannot be disposed of by a third party because of its chemical properties, it should be destroyed in the laboratory by a non-hazardous method, or should be converted into a disposable form. Special operating instructions should be drawn up for this. With regard to the use of waste air filters it should be ensured that the suppliers guarantee to take back the filter and correctly reactivate it or dispose of the filter medium in order to avoid waste production.

The collection and storage of waste within a plant should be carried out according to § 19 g, paragraph 1 of the Federal Water Act 'storage and transfer of substances hazardous to water'. Different types of waste should be collected separately in the laboratory so that dangerous reactions and hazards to personnel are avoided. The containers used must be suitable for collecting a particular type of waste with regard to size, construction and material, and must be able to be transported safely. They must be labelled according to TRGS 201.

The following instructions must be adhered to:

– Cyanides must not be poured into the drainage system even in small quantities, but must be collected separately. In some cases they may be treated with soluble iron salts.
– Waste materials which can produce poisonous or readily flammable gases and fumes or which react with water (sodium, potassium, carbides, and phosphides) are to be collected in fire-resistant containers and disposed of separately.

– Flammable liquids should not be poured into the drainage system but collected separately. (Chlorohydrocarbons must be kept separate from other solvents.)
– Solutions containing heavy metals should not be disposed of in the drainage system. In exceptional cases small quantities are pretreated. They are detoxified by 'filtration' through granulated magnesium oxide/marble (1:2). This mixture can be placed in a broad shallow column (prepared, for example, by cutting a polythene bottle) which is fixed over the sink. The charged filter is disposed of as special waste.

1.7 Waste Water in the Laboratory

The following legal regulations should be adhered to for the discharge of laboratory waste water:

– The Federal Water Act: appendix to § 7a for direct dischargers,
– Ordinance on the Indirect Discharge of Wastewater,
– Municipal Wastewater Regulations and the German Wastewater Association (ATV) technical leaflet 115.

According to the Sewage Origin Ordinance which was valid until 1997, laboratory waste water is defined as that in which hazardous substances can be found. Purification of waste water including pretreatment is carried out according to the state of the art. From 01.04.97 *all* substances contained in waste water are to be removed according to the state of the art. The limiting values for discharge for direct dischargers (i.e. only a few large plants and their laboratory waste water) are defined in the Discharge Permit. Indirect dischargers must obey the Ordinance on the Indirect Discharge of Wastewater and the local waste water statutes.

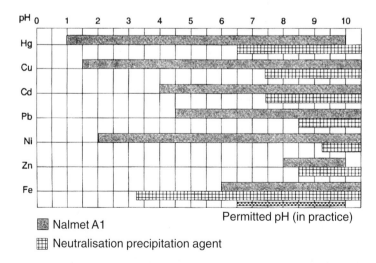

Nalmet A1

Neutralisation precipitation agent

Permitted pH (in practice)

Fig. 2: pH range for the precipitation of heavy metals in laboratory waste water

In large factories laboratory waste water is generally piped to the in-house waste water treatment plant. Smaller laboratories can send their waste water to an external agent for purification or neutralise it themselves. Disposal by outside companies is cost-intensive and has the disadvantage that the waste water must be collected and stored safely over a longer period. Neutralisation of the waste water at source lowers the risk to the environment and in most cases is sufficient for discharge into the public drainage system.

Small laboratory neutralisation plants with acid or alkali metering can treat more than 200 litres of waste water per hour and can be built into fume cupboards or laboratory benches. They are fully automatic, closed systems and guarantee a pH of between 6.5 and 9.0. Figure 2 shows the necessary pH range for the precipitation of heavy metals as the hydroxide or sulfide (with the usual organosulfur compounds such as TMT 15 from Degussa or Nalmet A 1 from Nalco).

1.8   Waste Air in the Laboratory

In spite of efforts to reduce the quantities of chemicals used in laboratories, there are processes in which the handling of hazardous substances is unavoidable. Examples are filling, separating and dissolving.

In laboratories where polluting chemicals, such as acids, bases and solvents, are used the pollutants liberated must be collected and removed using suitable techniques. In § 19(1) of the Hazardous Materials Ordinance, it says *'work should be carried out so that hazardous gases, fumes or suspended particles are not produced'* and in (2) *'if the measures taken according to paragraph (1) cannot prevent the production of these substances then they should be collected and disposed of at the point of production without posing a hazard to human beings and the environment, insofar as this is possible according to the state of the art'.*

The following laws and regulations must also be adhered to:

– Federal Immission Control Act,
– 4th regulation of the Federal Immission Control Act,
– Technical Instructions on Air Quality Control
– DIN 12924, section 2: laboratory equipment
– DIN 194, section 7: ventilation techniques
– Employment Accident Insurance Fund, chemical laboratory guidelines
– Maximal workplace concentration (MAK), Biologically Tolerable Workplace Concentration (BAT) and Technical Guide Concentration (TRK) values.

Laboratory air scrubbers can be used to purify polluted waste air. There are two variants: centralised and decentralised scrubbers. Centralised scrubbers generally employ filled columns for purifying waste air conducted from several different fume cupboards to a single location outside the laboratory. Decentralised scrubbers are predominantly sprays which are installed directly in the fume cupboard. Comparisons between both types have shown that decentralised scrubbers are better because of their greater safety and lower operating costs. Their use also fulfils the requirement of the Employment Accident Insurance Fund chemical laboratory guidelines that pollutants should be disposed of as close as possible to where they are formed or liberated.

While waste air scrubbers are predominantly used for the absorption of acids and alkalis, the solvents, odorous substances and pyrolysis products of incomplete combustion are preferably removed by adsorption on to activated charcoal. Various systems specially designed for laboratory use correspond to the state of the art.

Workplaces where hazardous substances are handled are increasingly moving from the laboratory bench to digesters or specially extracted areas such as chemical filling stations. There a variety of substances can be liberated, whereby it is not possible to use a universal purification system. Some substances can be easily removed with water whereas it is better to retain others on solid filters.

Waste air filters can now be integrated into normal laboratory equipment. Filter modules are incorporated into the upper parts of digesters or in neighbouring filter units. To achieve effective air purification the filter must be appropriate for the particular type of pollutant. Particularly in laboratories where shift work is carried out and there is therefore a frequent changeover of laboratory personnel, automatic controls are a good idea. The capacity of the filter module should be large enough to avoid frequent changing of the filter. Service lives of ca. 6 months should be aimed at.

1.9 Eco-Audits for Laboratories

Since April 1995 commercial enterprises, including laboratories, in EC countries can undergo a voluntary eco-audit, based on the eco-audit regulation of July 1993. The aims of the audit are:

– setting up an environmental management system,
– self monitoring by regular audits,
– external communication through an environmental declaration,
– external monitoring by environmental assessors.

Auditing involves the following steps:

– environmental testing,
– building up equipment for environmental protection,
– testing operations for their effect on the environment,
– an environmental declaration,
– certification.

The first environmental test is the recording of the actual state and has the aim of creating a secure database for all environmentally relevant processes in the laboratory. Technical equipment, material balances (e.g. waste streams) and available authorisations are recorded, subdivided into the areas of water, soil and air and the records drawn up as existing documentation. The functions of the existing building and waste disposal organisations are ascertained and analysed.

On this basis equipment for environmental protection is developed. Besides technical and product-based aspects, the overall organisation is included. The documentation regarding the equipment is produced in the form of an environmental handbook.

In the environmental operating test, which is the actual audit, the environmental protection equipment is tested and evaluated regularly and systematically. Environmental aims and the environmental programme of the company are brought up to date and measures for the further development of the environmental management system are proposed.

In an environmental declaration the general public is informed, for example, about the environmental protection aspects of the plant operations. A summary of the consumption of raw materials, energy and water, emission of pollutants and generation of waste supplement the data.

Certification of the company is carried out by an authorised independent environmental assessor. He checks the statements in the environmental declaration and the adherence to the instructions of the EC regulation. After successful certification the company receives the eco-audit symbol and is included in the list of audited companies. This is published once a year in the EC official gazette.

1.10   Industrial Medicine Aspects

According to the Hazardous Materials Ordinance, a laboratory operator is obliged to draw up operating instructions for handling hazardous substances, which describes possible hazards for human beings and the environment as well as the necessary protective measures. TRGS 555 and § 20 of the Hazardous Materials Ordinance should be taken into account.

Work Safety Plan
All instructions regarding protection from hazardous substances or micro-organisms are summarised in a work safety plan and are communicated to the personnel coming into contact with these substances both verbally and in writing. The safety plan includes the operating instructions which are obligatory according to § 20 of the Hazardous Materials Ordinance and is therefore partly a legal document. The work safety plan should include at least the following:

– name of the person responsible (laboratory manager, safety officer),
– proof of qualification of the employees (medical certification, certification of courses on respiratory protection and first aid),
– plans of the plant (laboratory, chemicals, solvents and waste stores),
– definition of particularly hazardous areas,
– nature of workplace measurements,
– sampling,
– planning for emergencies and first aid.

Measurements at the Workplace
Measurements of the MAK and TRK values of hazardous substances should be carried out at exposed workplaces in larger laboratories within the framework of a working area analysis. They include the listing of the hazardous substances and the specific limit values, the localisation of the danger area, possible exposure and the laying down of methods for control measurements. The MAK value is the maximum permitted concentration of a substance at the workplace (gas, vapour or suspended particles) which generally does not impair the health of or unreasonably affect the employ-

ees in general even on repeated and long term exposure (usually 8 hours per day during an average 40 h working week) (TRGS 900). In Germany MAK values are determined according to TRGS 402 (individual substances) and TRGS 403 (mixtures of substances). Gas measurements can be carried out for many individual substances with an absorption tube directly at the workplace. An overall analysis of mixtures of unknown substances can be carried out initially using a polytest absorption tube or with portable instruments with photoionisation (PID) or flame ionisation (FID) detectors. This can be supplemented by analysis of the individual substances.

Industrial Medical Examinations

According to Employment Accident Insurance Fund regulation 100 (VBG 100) a company must ensure that the employees are monitored during dangerous activities and/or effects by precautionary industrial medical examinations. These take place when the work is begun and are repeated at intervals which depend on the substance being handled. Examinations must agree with the Employment Accident Insurance Fund basis for precautionary industrial medical examinations (G4, G8, G15, G16, G29, G33, G36, G38, G40), depending on the potential hazard.

   Particular industrial medical examinations must be provided for the handling of carcinogenic substances. Here the following regulations must be adhered to: VBG 100; the Hazardous Materials Ordinance, appendices II and V; TRGS 102; TRGS 900, appendix IIIA; TRGS 905, (the bases for industrial medicine); and TRGA/TRGS (the initiation thresholds for carcinogenic substances). The initiation thresholds given in the TRGS/TRGA, which make industrial medical examinations obligatory, are to be adhered to in particular.

2 Quality control

2.1 General

Quality control measures have always been part of analytical work. However, the systematization of quality control has only recently been developed. The basic principle is that measurements should be as comprehensible, as plausible and as accurate as possible, or at least interpretable.

> Only those analysis results having verifiable accuracy and precision can be compared.

Measurements on 'identical' samples frequently differ. Therefore all external influences on the results must be taken into consideration to avoid significant errors in the individual steps of test procedures.

Figure 3 illustrates the problems encountered in judging and checking errors of measurement. Two types of errors can be identified: systematic errors and random errors.

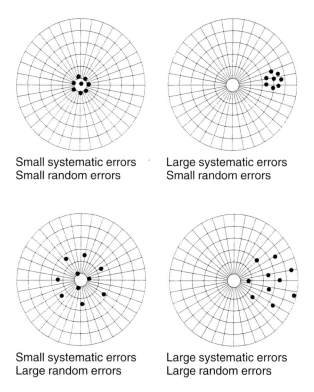

Small systematic errors   Large systematic errors
Small random errors   Small random errors

Small systematic errors   Large systematic errors
Large random errors   Large random errors

Fig. 3: Types of error in laboratory measurements

Analytical quality assurance is important for obtaining comprehensible and accurate results. The term covers all the procedures carried out: quality management, quality planning, quality guidance, and quality testing. This means that all the steps in a test procedure listed below form part of the analytical quality assurance system:

– sampling,
– sample conservation,
– sample transportation,
– sample storage,
– sample preparation/processing,
– measurement,
– data evaluation,
– test report.

2.2   Implementation of Analytical Quality Assurance

Implementation of analytical quality assurance involves the following (Landesamt für Wasser und Abfall, 1990) (regional office for water and waste, 1990):

– optimisation of personal equipment,
– optimisation of technical equipment,
– choice of suitable test procedures for the problem,
– determination of the parameters of the test procedure used,
– carrying out internal quality assurance procedures,
– participation in external quality assurance procedures,
– evaluation and documentation of the quality assurance procedures used.

(For accreditation, which is also part of analytical quality assurance, see Section 2.3).

Optimisation of Personal Equipment
The following prerequisites should be fulfilled when setting up a test:

– employment of suitably qualified personnel,
– the supervisor of the test should be qualified at university level,
– all co-workers performing the test should undergo an induction with regard to their tasks and obligations,
– possibilities for further training.

Optimisation of Technical Equipment
The technical equipment for a test should be suited to the problem to be solved and should be of satisfactory quality. Particular attention should be paid to ease of handling, susceptibility to faults, breadth of application, and sensitivity. In addition any appropriate construction modifications should be made and waste disposal in accordance with the regulations should be ensured.

Choice of a Suitable Test Procedure for the Problem

It should be determined whether the test procedure selected is suitable for the problem and is sufficiently sensitive. The maximum tolerable total error of the process should be estimated. This consists of the random error (which affects the precision of the process) and the systematic error (which affects the accuracy) (see Fig. 3). If possible, personnel should have some experience of the test procedure; otherwise an initial development phase must be incorporated.

Determination of the Test Parameters of the Test Procedure Used

During the development and routine phases the parameters both for the standard sample with a known content and for the actual sample with an unknown content should be determined. The processes used, including possible modifications, should be recorded together with the calculated parameters, i.e. *recovery rate, standard deviation of reproducibility, coefficient of variation of reproducibility, standard deviation of repeatability and coefficient of variation of repeatability.* The definition of these parameters can be found in DIN 38402, part 42. Many methods in analytical chemistry require calibration procedures; i.e. a calibration function must be calculated to enable concentrations to be derived from the measurements made. In the simplest cases the values of the concentrations of the standard solutions and the corresponding measurements can be plotted graphically. The calibration curve can then be fitted by eye. This process is usually not exact enough, so it is advisable to calculate the calibration curve using a linear regression. The calibration function:

$$y = a + b \cdot x$$

where *a* is the calculated zero value and *b* is the gradient of the calibration function (representing the sensitivity of the measurement technique), depicts the smallest possible deviation of all the data points from the straight line function. The linear regression model is based on three assumptions:

- linearity over a wide range,
- constancy of variation of the measured values over the whole range,
- normal distribution of data.

The actual calibration function lies within a range of reliability (confidence interval VB) (Fig. 4). The confidence interval depends on the deviation of the calibration points from the curve (residual standard deviation s_y) and the gradient *b* of the calibration function. A parameter which clearly shows the quality of a calibration function is the methodological standard deviation s_{x0}, which is calculated from

$$s_{x0} = s_y / b$$

The methodological standard deviation may be employed to compare different analytical procedures over the same working range with the same numbers and positions of calibration points. Figure 5 shows two possibilities for the determination of nitrite.

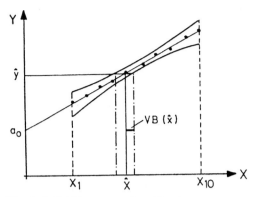

Fig. 4: Reliability range of a calibration curve

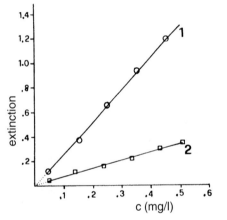

(1) sulfanilamide +
    N-(1-naphthyl)-ethylenediamine
(2) 4-aminosalicylic acid +
    1-naphthol

Fig. 5: Photometric determination of nitrite

It is clear that the residual standard deviations s_y are similar but the sensitivities differ. A numerical comparison of both s_y values using the F-test (see statistical references) shows that the two procedures differ significantly.

It must be remembered that every standard solution must be investigated using the same analytical procedure (including digestion where necessary) as the actual sample. As a minimum requirement for background quality control it is recommended that single or duplicate laboratory determinations are carried out on several consecutive days. These should include determination of:

– blank value,
– standard solutions with high and low concentrations within the working range,
– real samples,
– real samples after spiking.

Observations should be recorded over a period of 10 to 20 days. The stability of the system is then tested by calculating:

- s_w standard deviation within a batch,
- s_b standard deviation between batches.

The values of s_w and s_b are then compared using the F-test at a predetermined significance level (e.g. 95%). Only when this test reveals a significant difference between s_w and s_b must a separate calibration be carried out for each batch.

Internal Quality Assurance Procedures

Internal quality assurance involves carrying out procedures (daily or batch-related) which serve to recognise, eliminate and prevent errors in the use of a particular test procedure or measuring technique.

One method involves the use of control charts:

- mean value control chart for checking precision and accuracy,
- recovery rate control chart for checking matrix effects,
- range control charts for checking mean value distribution,
- difference charts for determining the deviation of experimental values from the mean value,
- standard deviation charts for monitoring the distribution of all the experimental values.

The different control chart systems should be drawn up using the company's own certified standards (so-called matrix standards or synthetically produced standards).

The use of control charts is based on the assumption that the errors in experimental data have a normal distribution. For each control sample (blank, standard or real) the total mean value $\bar{x}$ and the standard deviation s are calculated for the mean value control chart. The control chart is then constructed with the total mean value as its centre line. Warning and control areas are then added at distances of $\pm 2\,s$ and $\pm 3\,s$ respectively (see Fig. 6a).

Normally the control chart is followed over a period of one month. If all the measured values lie within $\pm 2\,s$, the analytical procedure is judged to be under control. The following errors are possible when the procedure is out of control (Fig. 6b):

1   values outside the warning range reveal gross analytical errors (e.g. errors in the preparation of standards, presence of impurities or incorrect calibration),
2   at least seven sequential results all lying above or below the mean (e.g. use of new standards of unequal quality),
3   at least seven sequential increasing or decreasing values (e.g. degradation of the standard through ageing or evaporation of the solvent),
4   sudden increased error (e.g. inadequacy of the method or insufficiently experienced personnel).

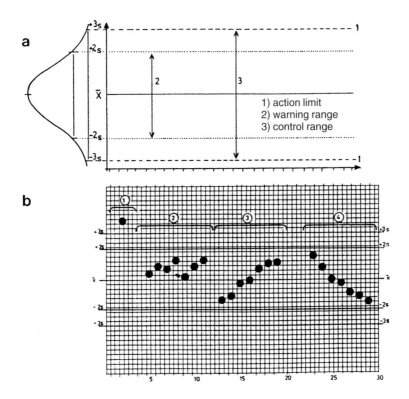

Fig. 6: Control charts   a Working ranges, b Detectable errors (see text)

Such control charts can also be used as recovery rate control charts (for the limited accuracy control of spiked samples).

A further possibility is the use of range control charts in which the differences between daily duplicate determinations are plotted. This is recommended for controlling the precision of measurements carried out on real samples when it is expected that significant matrix effects will alter the results. The control range of such a chart can, for example, be chosen to be 3.5 times the value of the mean variation from a previous period.

Participation in External Quality Assurance Procedures
Regular participation in national and international collaborative tests with synthetic samples of known content or with real samples of unknown content is important. With the latter matrix-dependent errors must always be reckoned with. The results of collaborative tests are usually recorded by the collaborative test managers and made available to the participants, often with appropriate recommendations.

Evaluation and Documentation of Quality Assurance Procedures Used
The quality assurance procedures used with the test should be recorded in detail (e.g. in an analytical quality assurance handbook). These records should be made available at the request of inte-

rested parties (clients, national monitoring organisations, or other departments within a large company). Insufficiently documented quality assurance procedures or those only carried out sporadically are of little value.

Problems of Analytical Quality Assurance Systems
The use and documentation of quality assurance procedures are indispensable for tests, in order to ensure that test results are really valuable, comprehensible, and justifiable. However, methods of quality assurance should be regarded realistically, as even with accredited or GLP-conforming tests, errors or inconsistencies can occur. The written formulation of quality assurance criteria by a laboratory is still not a sufficient guarantee of the quality of work. In case of doubt the customer can only instruct the company to use several different tests on an identical sample (test of accuracy). In addition identical samples from the same source can be tested several times under different designations (test of the standard deviation of repeatability).

The use of and adherence to quality assurance procedures usually increases the costs of tests. Whether this leads to better quality and more comprehensible and more comparable results depends on the individual case and the reliability of the test.

2.3   Statistical Tests

Where repeated analyses are carried out on the same homogeneous sample, identical analytical results are not obtained each time but a variety of results whose distribution can be represented by a histogram. If a large number of such measurements are available, a distribution curve can be constructed, which is often a normal Gaussian curve (Fig. 7).

This curve represents the relationship between the numerical value of an analytical result and its probability of occurrence. The distribution of the probability about its maximum is symmetrical. The main parameters describing such a distribution are the mean, $\bar{x}$ and the standard deviation s. Both parameters are particularly important for the evaluation of experimental data. The mean $\bar{x}$ is defined as

$$\bar{x} = \frac{x_1 + x_2 + \dots x_N}{N}$$

Where $x_1, x_2 \dots x_N$ are the experimental values.
The standard deviation s is given by:

$$s = \pm \sqrt{\sum \frac{\left(f_1^2 + f_2^2 \dots f_N^2\right)}{N-1}}$$

where f is the deviation of a single measurement from the mean and
N is the number of single measurements.

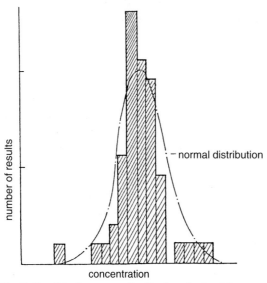

Fig. 7: Possible frequency distribution obtained from repeated analyses of the same sample

The standard deviation *s* is a statistical measure of the range into which the measurements of a series fall. At normal distribution

– ca. 68% of the values lie within the standard deviation, i.e. in the range $\bar{x} \pm s$,
– ca. 95% of the values lie within twice the standard deviation, i.e. in the range $\bar{x} \pm 2s$,
– ca. 99.7% of the values lie within three times the standard deviation, i.e. $\bar{x} \pm 3s$.

These values are obtained by integrating the corresponding areas under the normal distribution curve shown in Fig. 8.

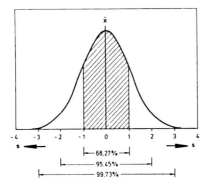

Fig. 8: Areas corresponding to 1–3 standard deviations (normal distribution assumed)

Sometimes it is more sensible to give the standard deviation *s* relative to the mean $\bar{x}$. This relative standard deviation is also known as the relative error or the coefficient of variation *V*. It is calculated according to:

$$V = \frac{s}{x} \cdot 100\%$$

Data are not always distributed normally. Other distribution forms can occur, particularly in evaluation of time-dependent water source data (e.g. river water analyses including flood water samples) or in measurement of the same sample by different laboratories during collaborative tests (Fig. 9). A coefficient of variation *V* of > 100% indicates data which does not fit a normal distribution.

Such results give an indication of noncompatible statistical populations (e.g. various water types) or of systematic errors of measurement. In these cases the different data sets must be statistically examined separately, since mean values and standard deviations will otherwise lead to incorrect conclusions. Such a case is shown in Fig. 10.

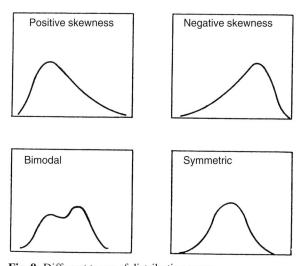

Fig. 9: Different types of distribution curve

Here the total number of measurements consists of two populations, since after about 1981 significantly higher values were found than before. The continuous curve represents the mean values, each calculated from five single measurements. A single statistical treatment of all values would be incorrect.

Trends also often lead to problems in interpreting data. In Fig. 11 graphs (a) and (b) have the same mean and standard deviation, although the measurements in (a) exhibit a periodical variation and in (b) a trend. It is therefore clear that additional aids, such as graphs and other statistical evaluation methods, cannot be dispensed with.

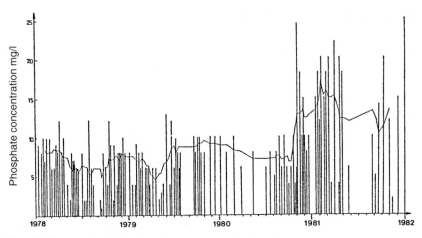

Fig. 10: Time series of the phosphate concentration in a river

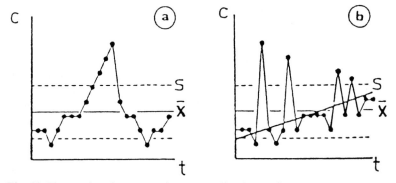

Fig. 11: Time series of concentrations (a) cyclic, (b) trend

Before experimental data are processed, extreme values or outliers must be identified and, if necessary, eliminated since most statistical techniques require a normal distribution. If this is not done a threshold can appear to be exceeded because of one analysis result that is too high, which is actually not the case. Statistical tests can provide objective clarification of such situations. Here test parameters are calculated using established equations and can then be compared with values published in tables. Important test methods, such as the F-test, the t-test, and the outlier test are described in specialised statistics literature. For some time efficient and PC-compatible statistics packages have been commercially available and permit the processing of all types of data. These can be stored in a format belonging to the same system or exported to other formats, such as Excel or dBase. Other formats allow the export of data for other applications. As well as a large number of simple and varied statistical processes, graphical representations are usually also possible. Such complex statistics packages include SPSS, STATISTICA, SYSTAT and UNISTAT.

2.4   Comparison of Different Quality Assurance Systems

There are currently three virtually independent quality assurance systems in Germany. To some extent their contents are identical, while other parts are specific to the particular system. The three systems are compared in Table 2.

The GLP system

The 1990 version of the Chemicals Act states that *'nonclinical experimental tests on substances or preparations, the results of which allow the evaluation of their possible hazards to human beings and the environment for the purpose of licensing, permission, registration, application or notification, must be carried out with adherence to the basic principles of good laboratory practice, according to this law'*.

The requirements of the basic principles of good laboratory practice (GLP) for laboratory tests are given in Appendix 1 of the law. Appendix 2 lays down the appearance of the GLP certification to be issued by the responsible regional authority.

Laboratories where substances are tested with regard to their hazards to human health and the environment must carry out their tests according to the basic principles of GLP. The data are recognised by the national authorities responsible, so duplicate tests in other countries can mostly be avoided.

GLP imposes specifications with regard to the organisation of a test (management, personnel), accommodation and equipment. It also prescribes a quality assurance system and lays down the necessary number of personnel, the measures to be carried out and the documentation of each step. GLP is thus conceived for laboratory work. It aims to make measurements comparable and, above all, comprehensible.

The EN 45000 system

The EN 45000 series of standards (including the DIN EN 45000 system) was drawn up by specialists in the European Community (EC). For test establishments, such as laboratories, they provide criteria in the form of an audit trail, which must be fulfilled in order to obtain a statement of technical competence through accreditation. These criteria are valid for the operation of the test laboratories (EN 45001), their assessment (EN 45002) and the authorities which accredit the test laboratories (EN 45003) and certify the products (EN 45011), the quality assurance systems (EN 45012) and the personnel (EN 45013).

The first step for a laboratory is the application to a licensed accreditation authority in order to fulfil the criteria of DIN EN 45001. The laboratory can only be accredited after being successfully examined by the accreditation authority. Part of the accreditation involves the occasional checking of the correct use of quality assurance procedures.

Compared with GLP, this system covers a wider area. The central feature of EN 45001 is the test laboratory. Areas covered include laboratory organisation, personnel, accommodation, equipment, operating methods and the quality assurance system used.

The ISO 9000 system

Companies, enterprises, organisations and not least local authorities produce material and non-

Table 2: Quality assurance systems (Oliveira, Praxishandbuch Laborleiter, 1996)

	Good laboratory practice	**Accreditation**	**Certification**
Law	National: Chemicals Act International: OECD Guidelines	DIN EN 45001 (1989) European standard	DIN EN ISO 9001–9003 (1994) International standards
Area of application	Stipulated for data concerning the safety of human beings and the environment on product licensing	Voluntary measure for all types of test laboratories	Voluntary measure for all production and service industries
Typical example	Toxicology or analytical laboratory attached to the research laboratories of a chemical company or a contract laboratory	Environmental analytical contract laboratory, (increasingly also all other types of test laboratories)	Analytical laboratory of a manufacturer which is part of the whole company
Goals	Comprehensibility through documentation, avoidance of test repetitions through recognition by international authorities, legal validity	Reduction in trade restrictions, comparability of test results, avoidance of test repetitions	Reduction in trade restrictions, creation of trust in the supplier and his products, improvement in quality in all areas of the company
Particular areas of emphasis	Organisational regulations and formalities, archiving, independence of the quality assurance unit	Precision and reliability of the results, calibration and validation of processes, internal and external quality assurance procedures	Internal and external interfaces, customer-supplier relationship, product design, corrective measures
Internal reasons for introduction	Cannot be circumvented during licensing of products, competition (contract laboratories)	Competition, improvement of quality, responsibility for products, management instrument	Competition, improvement of quality, responsibility for products, management instrument
Participating parties	Laboratory and monitoring authority	Test laboratory and contractor	Supplier and customer
Assessors	Appropriate regional authorities (inspectors)	Accreditation authorities, e.g. DACH, DAP, DAS-MIN, GAZ (assessors)	Certification authorities e.g. DQS, TÜV, DEKRA, LGA (auditors)
For whom is the licence valid?	Test establishment (laboratory) and 9 test categories	Test laboratory + test types/test processes/products (depending on the accreditation authority)	Companies or company divisions
For how long is the licence valid?	2 years (Germany)	Monitored annually (EC standard), review after maximum 5 years	Monitored annually, review after 3 years
Origin of the system	USA, toxicology	EC, with reference to ISO guide 25, important for the European internal market	International, important for the European internal market
Character of the system	Documentation system and in part quality management system	Proof of testing competence and quality management system	Proof of capability of quality work and quality management system
Motto	What has not been recorded has not been done!	Would I ask for a test to be carried out in this laboratory?	Is the supplier capable of quality work?

material products. For these products to be economical and of high quality a quality assurance system is required. The ISO 9000 series provides a framework, but it is designed to be so general that it can be applied to the widest spectrum of organisations. The central feature of this standard, which deals with quality management, elements of a quality assurance system, and stages in the proof of quality assurance, is the company and not the laboratory.

In order to remain competitive, companies aim to achieve certification which states that their development and production are carried out according to the quality requirements of ISO 9000. An independent authority tests this and may include the laboratory of a company, depending on the planned extent of the certification. In this case the quality assurance requirements stipulated in the standard must be fulfilled in addition by the laboratory.

If the laboratory wishes to carry out work for an external customer, it must fulfil the criteria of EN 45001.

3   Requirements for Analytical Methods

Before measuring programmes are begun the aims of the investigation must be laid down in order to be able to plan the work, which will include sampling, sample preparation and analysis.

The aims can include:

– control of recommended and threshold values,
– determination of the extent of pollution in water and soil,
– definition and laying down of pollution limits for environmental media,
– drawing up instructions for removing existing pollution and preventing fresh pollution,
– documentation of improvements in environmental protection,
– assessing and recycling waste.

Using modern test procedures almost every substance can be determined in almost every medium. In descriptions of methods for testing substances, mixtures or individual compounds in water and soil, sample processing and analytical determinations are mostly dealt with under the same heading. The following questions should be answered before preparing for and carrying out measurements:

– How are the samples obtained?
– Which sampling and measuring programmes are absolutely necessary?
– Are screening or *in-situ* tests sufficient?
– Do total parameters give sufficient information or should individual substances be analysed?
– What proportion of the substance is relevant (the total quantity or the soluble portion)?
– In what concentration range does the substance to be determined lie?
– Should the test only determine if a threshold value has been exceeded?
– Which reliability range is aimed at in the measuring process?
– Which test methods are published?
– How should the data obtained be documented?

The following goals may be formulated for the media to be examined:

Water
– determination of the actual water quality (official water quality register),
– determination of local and temporal tendencies (monitoring programme, forecasting models),
– determination of sources of impurities,
– test of the suitability of the water for a defined purpose,
– planning of water technology processes.

Waste Water
– estimation of pollution in sewers and sewage treatment plants,
– determination of ecological damage on discharge into bodies of water,

–   planning and operation of waste water treatment plants,
–   measuring detrimental parameters for the levying of fees according to the Waste Water Charges Act,
–   preparation for changing production processes.

Soil

–   estimation of soil quality depending on the intended use,
–   optimisation of fertilisation programmes,
–   forecast of soil changes (e.g. salting, acidification),
–   effect of soil on ground water quality.

Table 3 gives a list of the test groups most commonly used for the assessment of water and soil.

Table 3: Test groups for the examination of water and soils

Test group	Comments
1. Sensory examination	Essential for all samples; simple to perform (at collection or in the laboratory)
2. Physico-chemical measurements	Essential for all samples; simple to perform (at collection or in the laboratory)
3. Group analyses	Frequently performed; provide information on the type of contamination (laboratory)
4. Cations, anions and undissociated substances as major components	Often performed to characterise the chemical composition; rapid tests are available (in part during collection, mostly in the laboratory)
5. Inorganic trace analyses	Less common; provide information on e.g. contamination; rapid tests available in some cases (laboratory)
6. Organic trace analyses	Less common; give information on e.g. organic contaminants; only a few rapid tests available (laboratory)
7. Biological parameters	Give information on the biological environment and hygienic properties (in part during collection, mostly in the laboratory)

3.1   Ground Water

Ground water is water which is situated in underground aquifers. Figure 12 shows how ground water is replenished from rainfall, including the contact zone between the percolating water and the soil matrix, the unsaturated zone and the ground water (unconfined and confined aquifers).

The upper soil layer is the zone of intensive plant and microbiological activity and intensive addition of material through atmospheric deposition and soil use. In this zone organically bound nitrogen is converted into nitrate. The percolating water often has a low pH and a higher concentration of organic materials (humic substances) and aluminium. In the layer below the upper surface organic materials can be degraded or adsorbed on to clay minerals. The aluminium concentration decreases and the pH rises. Because of this, the water dissolves carbon dioxide and the solubility of the minerals generally increases. In the actual unsaturated zone the water tends to be part of a three-phase equilibrium between carbon dioxide, percolating water and calcite, whereby the carbon dioxide concentration can be up to 10 vol%. In the capillary fringe there is mainly mixing in the horizontal and vertical directions caused by changing levels of ground water. In the aquifer, which can consist of several hydraulically separated zones, the final composition of the ground water is established through contact with the solid rock matrix. The water mainly moves horizontally and thus comes into contact with other areas of the solid rock matrix. Anaerobic zones can be formed (reduced ground water) in which, for example, nitrate is reduced to elemental nitrogen or sulfate to sulfide. In general the chemical composition of the ground water is determined by the enrichment of readily soluble components, while sparingly soluble components of the rock reach their saturation concentration.

The rate of percolation in the unsaturated zone varies widely depending on the soil type. The rate of flow of ground water in sands is 1 to 5 m per day, in gravel 6 to 10 m per day and in silts and clays (ground water dams) the rate is often as low as a few mm or cm per day.

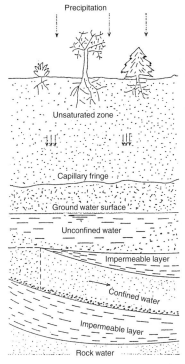

Fig. 12: Schematic representation of ground water replenishment and different types of ground water

The diverse nature of the rocky subsoil and the consequently varying composition of the ground water give rise to several important water categories, which can be characterised using a square diagram (Fig. 13).

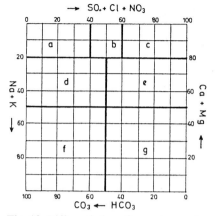

Normal alkaline earth water
a) Mainly hydrogen carbonate (bicarbonate)
b) Hydrogen carbonate-sulfate
c) Mainly sulfate

Alkaline earth water with greater alkali content
d) Mainly hydrogen carbonate
e) Mainly sulfate

Alkaline water
f) Mainly (hydrogen) carbonate
g) Mainly sulfate-chloride

Fig. 13: Differentiation of some types of ground water

Table 4 shows the relationship between the various concentration ranges of the most important dissolved components and the type of ground water.

Table 4: Concentrations of dissolved substances in different types of ground water, mg/l

Substance	Metamorphic rock	Sandstone	Carbonate rock	Gypsum rock	Salt rock
Na	5–15	3–30	2–100	10–40	up to 1000
K	0.2–1.5	0.2–5	up to 1	5–10	up to 100
Ca	4–30	5–40	40–90	up to 100	up to 1000
Mg	2–6	up to 30	10–50	up to 70	up to 1000
Fe	up to 3	0.1–5	up to 0,1	up to 0,1	up to 2
Cl	3–30	5–20	5–15	10–50	up to 1000
NO_3	0.5–5	0.5–10	1–20	10–40	up to 1000
HCO_3	10–60	2–25	150–300	50–200	up to 1000
SO_4	1–20	10–30	5–50	up to 1000	up to 1000
SiO_2	up to 40	10–20	3–8	10–30	up to 30

Contamination can occur, particularly in the upper ground water layers, as a result of the activities of human beings (refuse dumps, waste water seepage, spillages). Figure 14 shows how ground water is polluted by different forms of transport. Changes need to be assessed differently according to the use envisaged. For example, ground water may be used as crude water for drinking purposes, for industrial use, as a boiler feed or for irrigation. The measure of assessment of the quality of ground water has changed with time. For a long time there were limiting values given in the drinking water regulations. However, since 1994 the German LAWA (a working group for water set up by the German regional governments) has worked out a better basis for assessment in their 'Recommendations for the Recognition, Evaluation and Treatment of Ground Water Pollution.'

Table 5 lists specific parameters for the analysis of ground water.

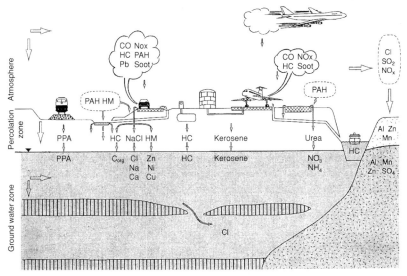

Fig. 14: Effect of transport on ground water quality (Golwer, 1995)

3.2   Surface Water

The overall ecological state of a body of water results from the interrelated action of all the biotic and abiotic influences. Water quality depends on many factors. Because of this, in limnology a combination of measured parameters and plant and animal indicators are often referred to. It seems sensible to base a quality scale on the natural state from which stages of increasing change can be described. However this sometimes gives rise to difficulties because in many cases it is almost impossible to reconstruct the natural state, or because a state where there is very little human interference is in need of protection.

This means that water quality is relative; it does not exist in isolation but describes the suitability of a body of water for a particular purpose. One possible definition is as follows: If it is assumed that the state of surface water can be characterised by concentrations or intensities of physical, chemical and biological parameters, and its suitability for a particular purpose can be determined from appropriate measurements, then the quality of the water is the state of the water assessed according to the requirements for use and described through various indicators.

Table 5: List of specific parameters for the analysis of ground water (GW)

Determination	General GW examination	GW contamination a) review	GW contamination b) comprehensive	Tests for concrete corrosion
1. Sensory examination				
odour, colour, turbidity	x	x	x	x
2. Physico-chemical				
temperature	x	x	x	
pH	x	x	x	x
electrical conductivity	x	x	x	
redox potential	x	x	x	
solids			x	
absorption at 254 nm	x			
3. Group analysis				
residue on evaporation			x	
ash			x	
oxidisability	x	x	x	x
DOC (dissolved organic carbon)			x	
phenol index			x	
hardness (Σ alkaline earths)	x			x
toxicity			x	
4. Cations, anions, undissociated substances				
Na	x		x	
K	x		x	
NH_4	x	x	x	x
Ca	x	x	x	x
Mg	x	x	x	x
Fe (total)	x		x	
Fe (II)	x		x	
Mn	x		x	
HCO_3	x	x	x	x
Cl	x	x	x	x
NO_3	x	x	x	x
NO_2			x	
F	x		x	
CN			x	
SO_4	x	x	x	x
S (sulfide)			x	x
PO_4	x		x	

Table 5: (Continued)

Determination	General GW examination	GW contamination a) review	GW contamination b) comprehensive	Tests for concrete corrosion
SiO_2	X		X	
O_2	X	X	X	
aggressive CO_2	X			X
5. Inorganic trace elements				
As			X	
B			X	
Cd			X	
Cr			X	
Cu			X	
Hg			X	
Ni			X	
Pb			X	
Zn			X	
6. Organic trace substances				
hydrocarbons			X	
volatile halogenated hydrocarbons	X		X	
involatile halogenated hydrocarbons			X	
polycyclic hydrocarbons			X	
pesticides			X	
7. Biological parameters				
colony count	X		X	
faecal indicators	X		X	
multicellular organisms			X	

In a body of water there are organisms with varying nutrient requirements in a relationship with each other and with the substances in their environment. In principle, an environmentally damaging chemical can affect the water in two ways: by providing nutrients for organisms or through direct toxic effects (primary effects). Depending on the situation of the organism in the water, consequences arise as a result of interaction with organisms of the same species, food sources, competitors and predators (secondary effects). Ultimately these lead to structural changes in the symbiosis (e.g. changes in the relative occurrence of organisms or shifts in the spectrum of the different species).

In choosing conventional parameters to evaluate water quality the PRATI classification can be used (Table 6).

Table 6: Assignment of water quality according to PRATI

Parameter	Quality			
	Very good	**Good**	**Slightly contaminated**	**Highly contaminated**
pH	6.5–8.5	6.0–8.4	5.0–9.0	3.9–10.1
Oxygen, %	88–112	75–125	50–150	20–200
Turbidity, mg/l	up to 20	20–40	40–100	100–275
Ammonium, mg/l	up to 0.1	0.1–0.3	0.3–0.9	0.9–2.7

LAWA uses chemical and biological assessment criteria as follows:

– *Quality Class I:* very low to low contamination
  Zones of pure water, always almost completely saturated with oxygen and low in nutrients; few bacteria; moderately densely populated predominantly by algae, moss, turbellarians and insect larvae; spawning ground of salmonides; BOD_5: < 1 mg/l; NH_4-N: trace; O_2: > 8 mg/l.
– *Quality Class I–II:* low contamination
  Zones with low inorganic or organic nutrient input without significant oxygen depletion; densely populated mostly with a large variety of species; salmonides present; BOD_5: < 1–2 mg/l; NH_4-N: ca. 0.1 mg/l; O_2: > 8mg/l.
– *Quality Class II:* moderately contaminated
  Zones with moderate contamination and good oxygen supply; very large variety of species and density of algae, snails, small crabs and insect larvae; abundance of water plants; large fish population; BOD_5: < 2–6 mg/l; NH_4-N: < 0.3 mg/l; O_2: > 6 mg/l.
– *Quality Class II–III:* critically contaminated
  Zones where contamination with organic, water-depleting substances has reached a critical state; fish may die through insufficient oxygen; reduction in the number of species of macro-organisms; certain species tend to increase in population; abundance of algae; on the whole fish population still high; BOD_5: < 5–10 mg/l; NH_4-N: < 1 mg/l; O_2: > 4 mg/l.
– *Quality Class III:* heavily polluted
  Zones with heavy organic, oxygen-depleting pollution and usually low oxygen content; local sludge deposits, colonies of rod-shaped waste water bacteria and fixed ciliates cover the surface and inhibit the occurrence of algae and higher plants; only a few insensitive animal organisms, such as sponges, leeches and water lice to be found in large numbers because of the lack of oxygen; low fish population; periodic fish mortality possible; BOD_5: < 7–13 mg/l; NH_4-N: 0.5–several mg/l; O_2: > 2 mg/l.
– *Quality Class III–IV:* very heavily polluted
  Only bacteria, fungi and flagellates; no higher organisms on a permanent basis; BOD_5: < 10–20 mg/l; NH_4-N: several mg/l; O_2: > 2 mg/l.
– *Quality Class IV:* excessively polluted
  Putrefaction processes; only bacteria, fungi and flagellates; no higher organisms BOD_5: >15 mg/l; NH_4-N: several mg/l; O_2: > 2 mg/l.

In the biological assessment of bodies of water, there is a significant connection between typical key organisms and the water quality. It should be noted that these key organisms are not limited to one water quality class, but there are overlaps between individual classes.

The producers (bacteria, blue algae, algae and higher water plants) normally have sufficient quantities of light, carbon dioxide and water available. Their growth is mainly affected by the concentration of nitrogen and phosphorus compounds and trace elements. The occurrence of animals depends on adequate supplies of oxygen and nutrients and the chemical composition of the water, among other factors. It also strongly depends on the morphology of the river bank, the river bed and the strength of the current. Heterotrophic micro-organisms (those requiring complex nutrients) are found wherever there is degradable material.

Gravel stream beds are always more highly populated quantitatively and qualitatively than muddy areas. This is because stones can carry dead material (detritus), moss, or higher plants.

The porous area of sandy stream sediments generally occupies 45% of the total volume. The pore system is supplied with water and detritus from the stream. Detritus is the nutrient basis for the fauna living in the porous area. Because of the absence of light primary production does not take place. Detritus deposits are rich in bacteria and fungi. The pores must not be too small so that they can still absorb detritus. Very fine particles clog the pores and thus lead to a decrease in the number of organisms living in them.

Bacteria are mainly found in sediments. The bacterial counts mostly lie between three and four powers of ten above those in free-flowing water. They are dependent on the content of organic substances and on the particle size of the sediment; the smaller the particle size the higher the bacterial count.

50% of the particles must be 0.25 to 0.50 mm in size so that the interstitial water in the upper sand layer can circulate. Such sands are populated by ciliates, turbellarians and rotifers, for example. If the proportion of fine sand (< 0.25 mm) increases to above 30%, anaerobic conditions are created leading to a sharp decrease in the number of animal species. Only certain ciliates, nematodes, rotifers and tubifexes can then still live there.

Damage to the structure and function of an ecosystem starts with damage to organisms. A substance is acutely toxic if it causes the death of a significant proportion of the population. For a particular species toxicity is substance-dependent and for a given substance, species-specific.

With a substance of sublethal toxicity the life parameters of an organism are significantly impaired, e.g. development is slowed down and/or reproduction is delayed.

On severe spasmodic contamination of a body of water the concentrated pollutant initially destroys all the sensitive organisms partly or completely. Plants are generally more tolerant than animals. The number of plant-eating animals decreases and thus the number of their predators also declines.

The appearance of pathogenic organisms in surface water should be noted, in particular where the water is intended for use as drinking water or for bathing. The following micro-organisms are indicators of unhygienic conditions:

– total coliform bacteria,
– faecal coliform bacteria,
– faecal streptococci,
– salmonella,
– intestinal viruses.

These bacteria are pathogenic to humans and can reach surface waters via domestic drains or waste water from hospitals or abattoirs. Their optimum living temperature is ca. 37 °C and therefore they tend not to multiply in normal bodies of water. Their infectiousness, however, can be preserved by certain surviving forms. In normal biological sewage works these pathogens can never be completely eliminated so that there is always the danger that waste water can cause microbiological contamination of surface water.

3.3   Drinking Water

Being the most important material for human consumption, drinking water must be free of pathogenic organisms and must possess no properties detrimental to health. Contamination through human activity can spread infectious diseases, such as cholera, hepatitis, helminthiasis (worms) or typhoid. Therefore continuous microbiological examination of drinking water is necessary and is a legal obligation. In addition water should also be tested for harmful materials, such as heavy metals, cyanides, nitrates, phenols or pesticides. In Germany drinking water is mostly obtained from ground or surface water. Ground water usually contains high concentrations of alkaline earth cations and is thus harder than surface water. The latter is more affected by human activity, while filtrates from river banks and enriched ground water lie between the two.

Crude water for the preparation of drinking water can be obtained from the following sources:

– ground water,
– water filtered from the banks of overground rivers,
– spring water,
– ground water enriched by surface water or purified waste water,
– river water,
– natural lakes,
– reservoirs,
– desalinated sea water,
– rain water.

The scope of the preliminary examination must take into account the type of untreated water while the examination of treated drinking water is subject to prescribed testing. Monitoring usually starts with bodies of water, continues through the storage reservoir and the various processing and distribution steps and ends with the consumer. Figure 15 describes this procedure schematically.

The minimum quality requirements for drinking water and surface water designated as a drinking water source are listed in Chapter 7. The scope of the examination depends on the minimum legal requirements. The control of water treatment operations, however, generally follows instructions drawn up within the company. In routine operations group parameters replace analyses of individual substances provided that this does not lead to significant loss of information.

The rapid availability of analysis results is important so that, when necessary, measures to improve water quality can be taken without delay (e.g. sterilisation depending on the untreated water quality, or de-acidification by mixing untreated water from various sources.). Above all this requires:

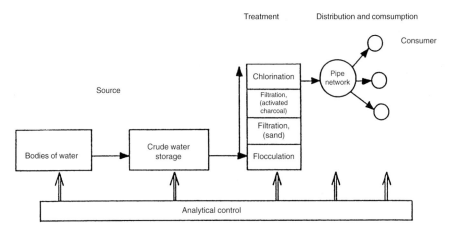

Fig. 15: Control areas in the treatment and distribution of drinking water

– a well functioning laboratory organisation including a sample management system,
– rapid and reliable sampling and testing procedures,
– well established analytical quality assurance, data documentation and interpretation systems,
– smooth co-operation with the water treatment plant.

3.4   Waste Water

Waste water is defined as that where the nature and composition have been changed through intensive human use. Waste waters can differ so greatly that neither a single classification nor a standard examination procedure are possible.

Waste water can pollute bodies of water through:

– readily biodegradable substances,
– non-readily biodegradable substances,
– plant nutrients,
– heavy metals,
– salts,
– waste heat,
– pathogens.

Reduction of the contamination by readily biodegradable, oxygen-depleting substances can only be achieved with mechanical/biological sewage treatment plants. The levels of the nutrients phosphorus and nitrogen are usually reduced through further waste water purification in specially equipped treatment plants. On the other hand, interfering and toxic substances, such as persistent organic substances, heavy metals and salts, should be removed at the point of discharge. Details of these procedures can be found in government regulations according to § 7a of the Federal Water Act.

Commercial and industrial waste water often contains substances which hinder the natural self-purification of bodies of water. These substances can be classified according to the negative effects which they cause:

- *Toxic substances* cause acute or chronic poisoning of water organisms.
- *Interfering substances* cause undesirable odours, tastes, colours, cloudiness or technical problems during processing, distribution and use.
- *Oxygen-depleting substances* reduce the oxygen content.
- *Nutrients* cause eutrophication of still or slow moving water.

Waste water can pollute bodies of water and give rise to far reaching consequences for drinking water supplies, fisheries, arable farming and animal husbandry. Before discharging waste water into the primary source it should be pretreated to reduce the pollution level and guarantee later use of the water. Continuous in-house and external control of waste water supports this procedure.

The EC supports the principle of pollution control through the monitoring of the discharge of waste water into bodies of water. This means that the requirements of the waste water treatment alone are made dependent on the usage and the state of the body of water. The connection between pollution and emissions remains variable from the ecological point of view. It depends on the natural conditions and can therefore not be forecast unambiguously. Also, since the future development of the nature and quantity of waste water discharges cannot be estimated with sufficient accuracy and the degree of purification depends on the available treatment techniques, in Germany, for example, the emission principle is predominantly in force. Waste water may only be discharged if the quantity and contamination are as low as is possible according to the generally recognised technical regulations or the state of the art. In the past there were minimum requirements regarding the discharge of waste water according to the government regulations given in § 7a of the Federal Water Act. A few years ago these regulations were revised as follows:

- Minimisation of quantities of water and pollutants is aimed at.
- The minimum requirements according to the state of the art are valid for waste water originating from areas where hazardous substances are used.
- Limitations in the concentration and loads specific to a production process are also possible in partial streams.
- The minimum requirements in force up till now are to be continued and made considerably stricter according to generally recognised regulations.

From 1997 only the state of the art applies to all discharges (transitional periods will be allowed).

Government regulations in Germany put emphasis on minimum requirements, i.e. the conditions imposed according to the water laws can and must take pollution aspects into account. For example, the nature and usage of bodies of water, the ratio of waste water to the lowest water flow, the water quality and, in certain cases, the protection targets of a particular area are decisive factors here.

Not all primary sources are used to the same extent and therefore do not all have to achieve the same quality. For different uses different qualities are achieved and guaranteed. For most types of use water quality standards should therefore be defined, but this has not yet been accomplished.

A single waste water characterisation is only possible with municipal waste water. Table 7 gives selected parameters for heavy, medium and low contamination.

Table 7: Average parameter concentrations in municipal waste waters with different degrees of contamination

Parameter	Degree of contamination		
	High	**Medium**	**Slight**
Total solids	1000	500	200
Sedimentable matter (ml/l)	12	8	4
Biochemical oxygen demand (BOD$_5$)	300	200	100
Chemical oxygen demand (COD)	800	600	400
Total N	85	50	25
NH$_4$-N	30	30	15
Cl	175	100	50
Alkalinity (as CaCO$_3$)	200	100	50
Oils and fats	40	20	0

The discharge of waste water components into sewage plants can inhibit biodegradation. Such damage may be expected to occur where the following concentrations are found (Table 8):

Table 8: Toxic concentrations of substances in sewage works

Toxic substance	Concentration (mg/l)
Cu	1–3
Cr(III)	10–20
Cr(IV)	2–10
Cd	3–10
Zn	3–20
Ni	2–10
Co	2–15
CN	0.3–2
H$_2$S	5–30

Table 9 illustrates how the extent of waste water examinations is established.

Table 9: List of parameters for the analysis of waste water

Parameter	Short analysis	Analysis to determine O_2 depletion	Comprehensive analysis
Colour, odour	X	X	X
pH value	X	X	X
Electrical conductivity	X		X
O_2	X	X	X
$KMnO_4$ consumption	X	X	
Chemical oxygen demand (COD)			X
Biochemical oxygen demand (BOD_5)	X	X	X
Total organic carbon (TOC),			
Dissolved organic carbon (DOC)			X
Na, K			X
Ca, Mg			X
Fe, Mn			X
Heavy metals			X
Total N			X
NH_4, NO_2, NO_3	X	X	X
Total P	X		X
SO_4	X		X
S (sulfide)		X	X
CN			X
Cl	X		X
HCO_3	X		X
Phenol index			X
AOX = adsorbable organic halogen			X
Surfactants			X
Putrefaction capacity	X	X	X
Fish toxicity			X
Toxicity to micro-organisms			X

3.5   Soil

Soil is the uppermost layer of the earth's surface and is formed as a result of weathering processes. Soil consists of mineral and organic materials (humus), generally with the addition of water, air and living organisms.

Soils provide the supporting environment for the higher plants and together they form an eco-system. Soil fertility depends upon the availability of water, heat and nutrients and the depth available for root growth. The latter is generally determined by the solum depth (upper soil layer). Where loose stones are incorporated into the subsoil, roots can penetrate deeper (Table 10). The root penetration depends on soil consistency, density, pore content, ground and standing water levels, pH and redox potential.

Table 10: Root penetration in soil

Depth (cm)	Root penetration
< 10	Very shallow
10–25	Shallow
25–50	Medium
50–100	Deep
> 100	Very deep

The aims of soil examination are as follows:

– determination of the nutrient availability,
– determination of the proportion of substances potentially damaging to ground water,
– determination of actual or potential salinisation,
– estimation of filtering efficiency.

An important goal of chemical soil analysis is the determination of mineral nutrients available to plants. This should take into account the depth of soil at the location (on sampling only the layer being analysed is usually taken into consideration) and the past and present weathering conditions. From the results the necessary quantity of fertiliser can be estimated.

The availability of individual nutrients (Table 11) is mainly dependent on the clay and humus content, soil humidity, pH value and redox potential.

Table 11: Availability of nutrients in soil

Degree of binding	Availability	To be determined
Unbound dissolved in soil	Very ready	Water-soluble nutrients
Partly bound to exchangers	Ready	Exchangeable nutrients
Immobile, readily mobilised	Moderate	Short-term reserve of nutrients
Immobile, not readily mobilised	Very slight	Total reserve of nutrients

The determination of inorganic contamination of soil is difficult since the natural content (background) is often unknown, depending on the rock from which it was formed, and the productivity and protective function of the soil are also important assessment criteria. The input of nutrients through fertilisation does not itself count as contamination, just as the removal of nutrients by plants cannot be considered to be decontamination.

The determination of mobile soil components involves considerable practical problems, e.g. in the case of heavy metals. The proportion of heavy metals available to plants should be eluted with an extractant, which allows a comparison with the heavy metal uptake by plant roots. Therefore mineral acids are as unsuitable as extractants as strong complexing agents (e.g. EDTA) are. Elution may be carried out with pyrophosphate or dithionite-citrate. The extractability of most heavy metals in soil is generally higher than that of most main group elements (phosphorus ca.1%, aluminium 2 to 3%, sodium 0.1 to 0.5%). However, it is known that heavy metals are more strongly bound in humus material than the elements mentioned above.

The tolerable limits of heavy metals in soil are important. Table 12 shows the toxicity thresholds. Above these concentrations impairment of plant growth can be expected. Limiting values, which must be adhered to when spraying sewage sludge on agricultural land, can be found in Chapter 7.

Table 12: Heavy metal toxicity towards plants in soil

Element	Toxicity threshold (mg/kg)
Cd	10–75
Cr	500–1500
Cu	200–400
Hg	10–1000
Ni	200–2000
Pb	500–1500
Zn	500–5000

Estimation of the corrosive effects of soil on metal pipelines and the determination of salinisation, especially in arid regions, constitute special areas of soil examination (see Section 6.3 and Chapter 7). Table 13 shows the scope of soil analysis as a function of the requirement.

Table 13: List of parameters for soil analysis

Parameter	Nutrient content	Toxic substances	Corrosion of metals	Salinisation
Particle size distribution	x		x	
Water content	x		x	x
pH value	x	x	x	
Redox potential		x	x	
Electrical conductivity	x	x	x	x
Acidity/alkalinity	x		x	x
Organic carbon	x		x	
Na, K	x			x
Ca, Mg	x		x	x
Mn	x			
Cu	x	x		
Zn	x	x		
Cd, Cr, Ni, Pb, Hg		x		
B	x	x		
Mo	x			
Total N	x			
NH_4	x			
NO_3	x			
Total P	x			
Cl			x	x
SO_4			x	x
CO_3, HCO_3				x
Sulfide S			x	
SAR value (sodium adsorption ratio) (see Section 6.3.2.8)				x

4 Organisation of Sample Collection Programmes and Sampling Techniques

4.1 General

The investigation of water, waste water and soil is normally aimed at

- quality control,
- making forecasts,
- determination of the extent of damage.

Most tests fall within the framework of quality control, and are frequently prescribed by legal regulations. For forecasts, data are determined for planning purposes or to recognise trends. This is important in the construction of sewage treatment plants, for example, where future discharge of waste water into bodies of water needs to be estimated. In estimating damage, the causes and extent of the damage are important.

The overall result of an analysis and the conclusions drawn from it cannot be better than the preceding sampling procedure. The most important prerequisite for proper sampling is the satisfactory qualification and training of the sampling personnel.

The most important goal is the collection of samples which are both representative and valid for the population to be tested. This means that they must be collected and stored in such a way that the parameters determined in the final sample correspond to the true values over the population of the water, waste water or soil. The location and time of sampling should be chosen in such a way that the samples reflect the temporal or local variance during the period of investigation.

Every sample collection depends to a certain extent on chance and is therefore subject to an inherent error. The smaller the sample, the less representative it is of the population. In addition, the information contained in the result from a random sample depends on the variation of the particular parameter measured. In order to be able to generalise empirical results, the size of the random sample error must be known. This error is the difference between the parameter of a random sample (e.g. arithmetic mean) and the real value for the whole population. The size of the random sample error depends on the sample population. Above a certain number of samples, the random sample error becomes so small that an increase in the number of samples can no longer be justified.

The variance in the overall analysis procedure, consisting of sampling, sample processing and analysis, is given by addition of individual variances, according to the error propagation law:

$$S^2_{\text{total}} = S^2_{\text{sampling}} + S^2_{\text{sample processing}} + S^2_{\text{analysis}}$$

With a sampling error of 25%, a processing error of 10% and an analysis error of 5%, there is a total error of 27%. If the processing and analysis errors are halved, the total error is only decreased by 2%.

In consideration of how representative the analysis results are, it can be seen that the most accurate measurement in the laboratory is of little importance when the errors in sampling and processing are considerably greater than the errors in measurement.

If the whole population of the material and the analytical methods are well known before sampling takes place, then a further lowering of the analytical error is not necessary if it accounts for ca.1/3 or less of the sampling error. This shows how important it is for the laboratory analyst to be responsible for the whole investigation including sampling.

An effective sampling programme takes the following into account:

– statistical aspects of the work,
– standardised instructions for sampling, and the labelling, transport and storage of samples,
– training of personnel in sampling techniques.

These requirements have been introduced in many areas of laboratory analysis as part of good laboratory practice (GLP) procedures. For sampling, these procedures do not yet exist.

Representative and valid samples can be taken in different ways for water and waste water (see the scheme shown in Table 14). Figure 16 shows three different types of sampling.

Table 14: Preliminary choice of sampling method

Concentration fluctuation	Flow variation	
	Small	**Large**
Small	(Qualified) random sample	(Qualified) random sample
Large	Time-dependent pooled sample	Volume- or flow-dependent pooled sample

For water and waste water, the following sampling procedures are used:

Random Sample
An individual sample is taken manually or automatically at a given time. It describes the state of the water at this time only.

Qualified Random Sample
This is a variant of the random sample where at least five random samples are taken at intervals of not less than 2 min over a maximum total period of 2 h and are then combined to give a pooled sample.

Time-dependent Sample
In the chosen sampling period samples of the same size are taken at the same time intervals and are combined to give a pooled sample. The result obtained is thus strongly dependent on changes in flow and pollution level of the water.

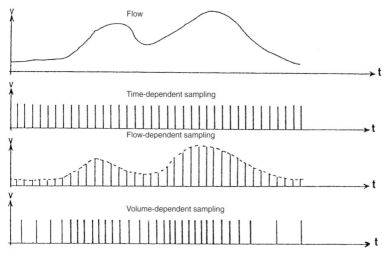

Fig. 16: Possible techniques for sampling water and waste water

Volume-dependent Sample

Here constant volumes are taken at variable time intervals determined by the volume of flow and are combined to give a pooled sample. If there are large variations in both the pollution level and flow (small dischargers) this sampling method should not be chosen.

Flow-dependent Sample

Samples are taken at uniform time intervals, which are of different sizes depending on the flow at that particular moment, and are combined to give a pooled sample. This type of sampling gives exact results even when there are large variations in flow and pollution level.

Incident-dependent Sample

This type of sampling is used if the exceeding of threshold values needs to be documented. Sampling only takes place when an incident has occurred; otherwise the sampler is only on stand-by. All continuously measurable parameters can be used to initiate sampling if a given measurement signal is exceeded.

Time-, Flow- and Incident-dependent Sample

This type of sampling is the most versatile combination of the different techniques. The priorities can be laid down, e.g. in the following order: incident-dependent, quantity-dependent, time-dependent. The process makes considerable demands on the sampling instrument and its control. A single instrument can deal with several tasks simultaneously, for example, continuous monitoring of a body of water (time- and volume-dependent sampling) at the same time as responding to incidents or faults (incident-dependent sampling).

Volume- or flow-dependent pooled samples can also be prepared manually afterwards, by combining several random samples consisting of different quantities of water (flow measurement necessary). Table 15 gives an example.

Table 15: Example of a manually collected quantity-dependent pooled sample

Time	Momentary flow (m³/sec)	Proportion of random sample in the pooled sample (l)
8	1	0.1
9	2	0.2
10	1.5	0.15
11	4	0.4
12	0,5	0.05
13	1	0.1
14	2	0.2
15	1.5	0.15
16	0.5	0.05
		Σ: 1.41

With soils, there is the danger that because of the granular nature of individual fractions, proportions are determined which are either too large or too small. The sample error S_S for a measured parameter is given as follows:

$$S_s = \frac{(1-x)}{x} \cdot \frac{\overline{m}}{m_s}$$

x content of component 1
$\overline{m}$ average mass of a particle
m_s mass of the sample.

The sampling error increases as the content of x decreases, with decreasing sampling area and with the increasing average mass of a particle. By increasing the number of samples, the total error will decrease more sharply than by making a larger number of parallel measurements. The conditions become more complex with mixtures of components with more than two different particle sizes.

4.2   Organisation of Sampling Networks

In many cases no reference can be made to an existing measurement network for the examination of water, waste water or soil, so a new network needs to be planned. The planning, including preliminary tests, needs to be carried out carefully, as the network will generally have to be used for a long period and subsequent modifications lead to difficulties in comparing analytical data. Since larger catchment areas or drainage systems of large cities consist of individual, often heterogeneous subsystems, measuring networks and sampling points must be planned taking this into account.

The main priority is that both should be representative and only be set up with regard to favourable transport connections and accessibility as a second priority.

For preliminary examinations, topographic and specialised maps (e.g. ground maps, geological or hydrological maps), aerial photographs and plans of bodies of water and drainage systems are used. If such aids are not available, provisional maps and plans must be drafted after inspection of the area. Further information is obtained by simple tests carried out in the field. In setting up sampling networks for bodies of water, hydrogeologists should be consulted.

Where samples of contaminated ground water are to be taken, there should be at least two measuring points. The first should be in the ground water exit stream of the contamination source. The distance of this point from the contamination should be < 10% of the flow distance of the ground water below the source. The second measuring point is in the feed channel of the contamination. The distance from this point to the contamination should be ca. 50% of the flow distance of the ground water below the source of contamination. The connecting line between the two measurement points should be perpendicular to the actual ground water level.

In order to examine the extent of pollution caused by old waste deposits more accurately, a sealed sampling network must be set up downstream. It is a good idea to set up measurement points perpendicular to the direction of ground water flow at regular intervals of not more than 50 m.

The diameter of a measurement point should be not less than 50 mm, if possible DN 125 or DN 150 in order to make sampling easier. With the latter normal submersible motor-driven pumps can be used. If the system is to be installed or removed frequently, there is enough space even if the pipes are not absolutely perpendicular. The diameter of the extension at the measurement point should allow a throughput rate of at least 0.1 l/sec without decreasing the ground water level too much.

The materials used at the measurement point must not be attacked by the contents of the ground water and should also not affect its chemical composition. Inert materials, such as special steel and fluorinated plastics, are suitable though expensive. For cost reasons rigid PVC is predominantly used in Germany to check the composition of ground water, in particular in cases of pollution by organic materials.

For river systems partial catchment areas must sometimes be taken into account when planning sampling networks (Fig. 17).

Here the total area can be divided into five sections, whose water quality is determined at a particular measurement point. Depending on the requirement, a further subdivision can be made. Critical areas in larger river systems lie below the junctions with strongly polluted tributaries and large waste water inflow points. In addition, less contaminated places should also be included in order to determine the natural composition of the water. Measurement points are often set up in rivers on both sides of a political border to resolve conflicts regarding water use. Generally one sampling point for every area of 100 to 200 km^2 is sufficient, except in industrial areas where more are required.

A section of the catchment area of a body of water is shown in Fig. 18. Measurement points at which possible changes in water quality might be expected are shown as circles. Populated areas have been shaded and waste water entry points (outlets from sewage plants and factories) are marked by arrows.

In addition to the major measurement points, regular checks can also be made at these entry points.

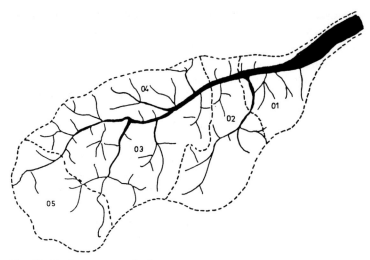

Fig. 17: Catchment area of a river system

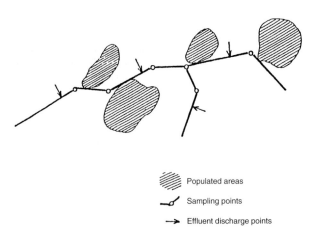

⬜ Populated areas

⟿ Sampling points

→ Effluent discharge points

Fig. 18: Part of a river system (schematic)

Figure 19 is a schematic representation of a network for waste water disposal in towns. It contains the street channels, collectors, main collector, collection shaft, overflow chamber and overflow channel as well as the sewage plant with an outlet into the body of water. Measurement points in the drainage system should be set up where individual industrial plants discharge waste water into the network, where main collectors meet and at sewage plant outlets. If a discharge of pollutants is suspected, the guilty party can be traced along his branch of the network.

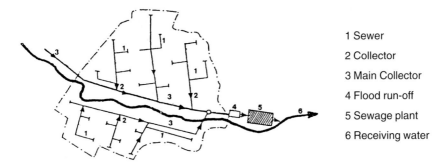

1 Sewer

2 Collector

3 Main Collector

4 Flood run-off

5 Sewage plant

6 Receiving water

Fig. 19: Scheme of a sewage network

   Sampling points in large rivers or water channels should be located such that waters entering above the sampling points are completely mixed. If this is not the case two sampling points can be positioned, one on each side of the waterway. Figure 20 shows how the mixing of waters from tributaries and waste water inlets is very slow when there is laminar flow. Here a permanent sampling point should not be placed before position 5.

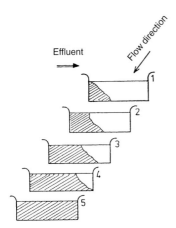

Fig. 20: Schematic representation of effluent mixing in flowing water systems

   In still bodies of water (lakes, reservoirs) with irregular shapes considerable heterogeneity in water quality is possible in the horizontal plane. It is therefore necessary to set up several sampling points, the number of which may possibly be reduced at a later stage. Heterogeneity may also occur in the vertical direction so sampling must also be carried out at different depths, e.g. near the surface in layers at different temperatures, and at the bottom (effect of sediment).

   In addition to the choice of a representative spatial sampling network, the choice of a suitable sampling timetable is also important. If concentration fluctuations are high, samples are taken more frequently, but if they are low, sampling at intervals of a few months can be sufficient for an assessment. Where cyclic variations in days, months or years are known, the sampling interval must be made more flexible in order to avoid repeated measurement of a high or low value (Fig. 21).

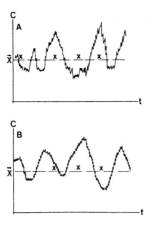

Fig. 21: Effect of concentration fluctuations on values predicted from random samples (A random, B cyclic)

After a longer test period, the most important statistical parameters, such as mean values, extreme values, variances and correlations should be calculated. Cyclic concentration fluctuations can easily be detected graphically. The sampling frequency can then often be decreased without loss of important information.

4.3   Determination of Water Quantities

The flow in rivers and non-pressure and normal pipelines can only be determined from indirect parameters. The quality of the measurement is thus dependent on the changes in these parameters over the whole area of measurement. While in pressure pipelines the cross-sectional area remains constant, in non-pressure pipelines and rivers both this and the average flow rate can vary.

Measurement of flow rate can only be carried out easily during sampling of ground and drinking water. In both cases the measurement is made either with a graduated container and stopwatch or by reading a water gauge within a defined time interval. With enclosed pipelines of different bore diameter flow measurements are usually carried out using magnetic-inductive methods.

In open channels or rivers measurement can only be carried out with a large degree of technical and mathematical effort. More exact determinations normally employ systems based on the Bernoulli flow equation. In hydraulic processes the cross section of an open channel is partially blocked so that an unambiguous connection between the flow and the water level immediately above the narrowing is established. Venturi channels or measuring weirs are generally used. Venturi channels have symmetrically arranged narrowing points on the sides of the cross section of the channel, while the base of the channel remains unblocked. Because of their importance, they have been standardised (DIN 19559, part 2). Measuring weirs with triangular, rectangular or trapezium-shaped notches are made from plates with fixed overfall weirs and are installed in open channels perpendicular to the flow direction.

In many rivers and canals the connection between water level and flow at a defined level indicator has been determined for several years and recorded graphically as flow curves. These are recorded by the regional authorities responsible in yearbooks.

Some well proven methods will be described below.

Open Channels

The Manning-Strickler method is frequently used for open channels. The channel gradients, wall roughness and hydraulic perimeter must be known. Where these three values are given, the quantity of flowing water can be simply determined by measuring the water level.

$$v = k \cdot R^{2/3} \cdot J^{1/2}$$

v velocity (m/sec)
k roughness ($m^{1/3}$/sec)
R hydraulic radius F/U (m)
J gradient (m/m)
F cross section (m^2)
U wetted perimeter (m)

The following standard k values are used:
90–135   for new asbestos cement pipes
85   for new stoneware pipes
65–75   for encrusted pipes

For right angled channels, R has the values

$$R = \frac{Width \cdot Water\ Depth}{(Width + 2 \cdot Water\ Depth)}$$

The flow in m^3/sec is given by:

$$Q = v \cdot F$$

Open Channels with Irregular Cross Section (Streams, Rivers)

a) Float

A float (e.g. a cork) is thrown into the middle of the flowing water and the time taken to cover a certain distance is measured.

The following equations are used:

$$v = \frac{l}{t}$$

$$Q = \frac{l}{t} \cdot F \cdot c$$

l measured distance, m
t measured time, sec
F cross section of flow, m^2
c coefficient

The flow cross section can be determined graphically or estimated by measuring the depth at various points and the width. For bodies of water which do not have dense undergrowth on the banks or coarse pebbles on the bottom the value of c is assumed to be 0.8 to 0.9. In other cases it is 0.5 to 0.8.

b) Measuring by Hydrometric Propeller
The propeller method is employed in those cases where larger flow cross sections are encountered and the precision requirements are more stringent. Calibrated propellers are used (Fig. 22).

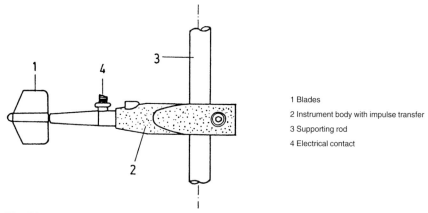

1 Blades
2 Instrument body with impulse transfer
3 Supporting rod
4 Electrical contact

Fig. 22: Hydrometic propeller for determining flow rate

The propeller is dipped into the water on vertical graduated plumb lines and the velocity is measured at these points at different depths. Less stringent requirements may be satisfied by taking measurements on each plumb line at 0.4 times the total depth. The procedure and the calculation of results are carried out as shown in Fig. 23.

A

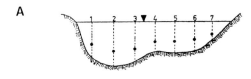

A Cross section of river showing plumb line
   measurement points

B Graphical determination of flow

B

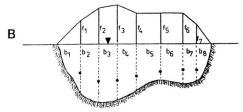

Fig. 23: Calculation of flow results using point measurements (Herrmann, 1977)

The measurements are made at the points shown at 0.4 times the total water depth. The velocity is multiplied by the water depth and plotted as f_i. The f_i points are then joined to form a polygon and the sum of the single areas formed from f_i and the single widths b_i are calculated. The flow Q (m³) is then calculated according to

$$0.5 \cdot \sum_{i=1}^{k}(f_i \cdot b_i + f_k \cdot b_{k+1})$$

Water gauge   Wear

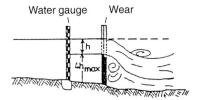

Wear crest

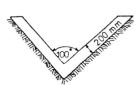

Fig. 24: Triangular wear according to Thompson

c) Measuring Weir

Measuring weirs are suitable for irregular flows not exceeding 1 m³/sec. Triangular measuring weirs are often employed. The construction is shown in Fig. 24. If the supply channel is triangular for 2 m like the weir itself and if the weir has the dimensions shown in the figure, Q (m³) is calculated as follows:

$$Q = \mu \cdot h^{5/2}$$

h the overfall height, m
μ 1.46 (at the dimensions shown)

d) *Salt Mixing Procedure*
This process is applicable to turbulent waters with rapid mixing. Introduction of foreign materials, such as salt, however, requires an official licence. A salt solution of known concentration is required, together with an electrical conductivity meter. The salt solution is allowed to flow into the water under defined conditions. The flow is then calculated from:

$$Q = A(B\text{-}C) \cdot (C\text{+}D)^{-1}$$

A flow of salt solution
B conductivity of salt solution
C conductivity of river water with salt solution
D conductivity of river water without salt solution

4.4   Sampling Devices

Sampling devices should possess the following properties:

–   They should be robust, suitable for use in the field, easily transported, handled and cleaned.
–   The sample should not be altered by reaction with the material from which the device is constructed.

Depending on the type of sample collection, simple apparatus or specialised equipment (e.g. pumps, automatic samplers) is employed. Other auxiliary equipment includes cool boxes, tubes, ropes, cables, plastic bags and spades.

The choice of sample container is of prime importance. Polyethylene or glass bottles are used in most cases. Samples containing nonpolar organic materials (e.g. oils, pesticides) should not be placed in plastic bottles, whereas glass bottles are unsuitable for samples in which low concentrations of sodium, potassium, boron or silicic acid are to be determined. Both types of bottle are suitable for higher concentrations of inorganic substances.

If contaminated samples are stored for long periods solids can settle on the container walls, so the containers should be washed thoroughly after use. Cleaning is first carried out mechanically and then with chromic acid. The latter must not be used with plastic containers. For these dilute hydrochloric acid is recommended. Stubborn impurities in plastic bottles are difficult to remove so these are discarded after pouring out the liquid. When using household detergents it should be noted that even after prolonged rinsing with water, surfactants and phosphates can still be released from the walls.

Sample containers for bacteriological examinations must be of glass and be sterilised before use by heating for a prolonged period at 180 °C together with their stoppers. The bottle necks are protected by aluminium foil.

Ground Water

Buckets are the simplest devices for the collection of water samples. They are mainly employed only for preliminary examination purposes or in those cases where the ground water mains are well mixed. They consist of cylindrical containers, like Ruttner buckets, and are thoroughly washed through with water when lowered into the well. At the sampling depth the bucket valves are closed through activation of a drop weight. The devices are suitable for defined sampling at any depth. Attention must be paid to the tightness of the seals. Such buckets are available in various diameters so that narrow bore holes can also be sampled. Buckets which are only opened on one side should only be used as auxiliaries on sampling from wells.

Sampling from wells or ground water measuring points is mainly carried out using pumps. With suction pumps the sampling depth is limited to 7 to 9 m, as in deeper waters the column of water breaks in the sampling tube. A valve at the end of the tube is necessary at depths of more than 3 m. Suction pumps can be powered by electric or petrol motors. The latter are unsuitable for samples where hydrocarbons are to be determined. At smaller pumping depths the delivery capacity of the pump is between 1 and 2 l/sec.

Submersible motor-driven pumps are centrifugal pumps connected to a submersible motor. Small pumps have a diameter of 95 mm and therefore can only be employed in bore holes of over 100 mm diameter. In the case of crooked bore holes the pump can be left hanging and thus lost. The pump is let into the well on a safety rope together with the sampling tube and the power cable. A generator is used to deliver electricity when no mains supply is available. The delivery capacity is to be found in the manufacturer's description. Submersible motor-driven pumps are mainly used in well mixed ground wells. They should not be used where there is a limited water supply as the water level can sink rapidly causing the pump to take in air. In special cases piston displacement, compressed air, or deep suction pumps (water jet principle) can be used. Underwater pumps of small bore diameter with low delivery capacity are mostly driven by car batteries.

All devices, such as pumps, cables and ropes, which come into contact with the water sample, must be designed so that no change in the water contents occurs. Metal components or lubricants for the pump can falsify the results of metal and hydrocarbon analyses.

The tube material should also not affect the sample. Polythene, silicone and PVC tubes can be used, provided that lipophilic organic substances, such as oils, solvents, pesticides and surfactants, are not being analysed. In these cases PTFE or metal tubes should be used.

Pumps and tubes should be cleaned and dried after sampling to prevent corrosion of metal components and the build-up of micro-organisms on plastic items.

Surface Water and Waste Water

Buckets are used as manual devices for the sampling of water from or near the surface and sealable ladling devices (Ruttner buckets) for deeper layers.

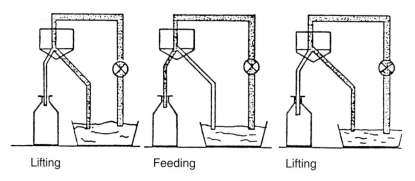

| Lifting | Feeding | Lifting |

Fig. 25: Schematic representation of an automatic free-fall sampler

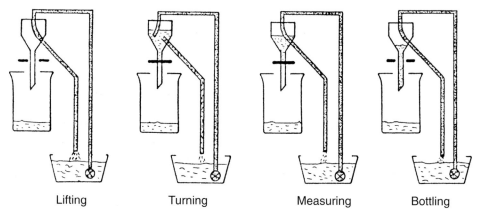

| Lifting | Turning | Measuring | Bottling |

Fig. 26: Schematic representation of an automatic free-fall sampler coupled with a fixed quantity feeding device

Automatic sampling devices are mainly used for waste water and consist of the following parts:

– delivery system,
– control mechanism,
– sample apportionment and metering,
– sample storage.

Delivery is accomplished using hose pumps or eccentric screw pumps. A system with a free falling water device is shown schematically in Fig. 25. In this way a sample is diverted from a continuous water stream using an impulse circuit (time- or quantity-dependent) so that a partial sample can be taken. This type of sampler is easy to use and clean. A cooling unit should be incorporated for longer sampling periods.

Figure 26 shows the sampling phase of a system within a free-falling water device coupled with a constant sample feed device. In this way, a constant sample volume is always obtained. Highly contaminated waste water can lead to blockage of the feeding container.

Short sealed pipes, high delivery velocities, and elimination of light are required to minimise changes in the nature of the sample. Regular cleaning and maintenance avoid deposit formation and ensure the reliability of the system.

Soil

Sampling of soil for the analysis of nutrients or pollutants or salt testing generally takes place at depths not exceeding 50 cm. Therefore a spade or simple boring device is normally sufficient. The latter can be a simple drill or a special drill head incorporating cut out walls (Pürckhauer drill), which is forced into the ground with the aid of a heavy mallet and removed by turning. The soil profile can then be seen in the drill grooves. Other necessary equipment includes a spatula or knife, a tape measure, sample bags or jars, a magnifying glass, and 10% hydrochloric acid to test for carbonates in the soil.

4.5   Preservation, Transport and Storage of Samples

The contents of water samples can alter at different rates. As only a few parameters can be measured during collection, pretreatment or stabilisation is often necessary. This allows tests to be carried out even after long periods of time have elapsed. Most inorganic components do not require additional measures for transport and storage, but some can undergo changes, for example through reduction, oxidation or precipitation. If the sample contains organic material and there are favourable conditions for the development of micro-organisms, rapid changes often occur. In these cases stabilisation is necessary. This is defined as the blocking or delay of biochemical processes whereby the difference in content between the original and preserved sample should be less than 10%.

Water and waste water samples are cooled during transport and storage, particularly at high external temperatures, as otherwise the rate of biochemical processes increases. Reactions generally take place in waste and surface water samples more quickly than in ground or drinking water samples.

The following changes are possible:

–   oxidation of components by dissolved oxygen (e.g. Fe^{2+}, S^{2-}),
–   precipitation and coprecipitation of inorganics through changes in the medium (calcium carbonate, metal hydroxides),
–   adsorption of dissolved trace components on the container walls,
–   changes in parameters as a result of microbiological activity (pH value, oxygen, carbon dioxide, biochemical oxygen demand, trace organics).

Some suitable preservation methods are listed in Table 16.

Table 16: Preservation of substances present in water

Parameter	Preservation method	Maximum storage time
Trace metals	5 ml HNO_3 per l	Several weeks
DOC, TOC, COD, BOD_5	Cooling at 4 °C or freezing at –18 °C	One day or several weeks respectively
NH_4, total N	5 ml HNO_3 per l	A few days
Hg	2 ml $HNO_3/K_2Cr_2O_7$ solution per l (0.5 g $K_2Cr_2O_7$ in 100 ml 30% HNO_3)	A few days
Cyanides	Basify to pH = 8	1 day
Fe(II)	Addition of 2,2'-bipyridine	1 day
S^{2-}	2 ml 10% Zn acetate solution per l	1 week
Phenols	5 ml 35% HCl + 1 g $CuSO_4 \cdot 5H_2O$ per l	1 week

These methods are only recommendations. Usually examination immediately after sampling makes preservation unnecessary. Particularly with relatively pure water, cooling to 4 °C is sufficient even for longer storage periods. If waste water has to be tested for COD or BOD_5 only after a long storage period, it can be frozen at ca. –18 °C in plastic bottles. Rapid freezing and thawing are important. Soil samples should be dried as quickly as possible, provided that tests in the original state are not necessary. For the testing of nitrate, the sample should be cooled during transport and then frozen in a plastic bag at –18 °C.

4.6   Sample Collection Procedure

4.6.1   Ground Water

Ground water samples are usually taken from ground water measuring points, bore holes or shafts with pumps or buckets, the use of pumps being preferred. Buckets should only be used in special cases, to remove pollutants from the upper surface of the groundwater or at the base of the aquifer. The disadvantage of buckets is the spreading of material on introduction to the deeper zones of the aquifer. The water taken then does not originate from the ground water pipe, but from the contents of the measuring point. In the case of bore holes the pump has to be lowered into the filter area. In measuring points with a sump pipe at the lower end samples can be falsified if mud, biomass or organic phases with density >1 become deposited in the sump pipe and do not get into the conveyed stream in reproducible amounts.

Depending on the aims of the investigation, sampling from ground water measuring points can be used to obtain flow-weighted or depth-orientated layer samples.

In flow-weighted pooled samples from thoroughly filtered measuring points a continuous flow is pumped out, from which samples are taken. Before sampling the stationary water in the filter pipe, the annulus and in the attached ground water pipe should be replaced. The DVWK leaflet 208 recommends replacement of 2 to 5 times the pipe volume, while according to the DVGW leaflet W 121, pumping should be continued until parameters which are easy to measure, such as pH, electrical conductivity and temperature become constant. The volume of the lifting tube, through which the sample is conveyed to the surface, should be replaced at least once. Where the ground water channels are in separate layers, the concentrations of the pooled samples can vary with time depending on the depth of sampling. For representative sampling, the sample volume should be as large as possible. It is therefore best to use a large storage container from which the analysis sample with a representative mixing concentration is taken.

To determine differences in the composition of the ground water over the whole thickness of the aquifer, depth-orientated ground water samples are taken. At continuously filtered measuring points sampling must take place by simultaneous pumping at different depths. To avoid short circuit currents within the measuring point each separate sampling area is separated hydraulically from its neighbours by packing. Co-ordination of the pumping rates at the various sampling depths with the permeability of the background at these points improves the quality of sampling. Using this process measuring points with partial filtering at the planned sampling depth are better than completely filtered measuring points.

The water level in the sampling point is measured using a light lead before and after sampling so that conclusions can be reached concerning the subsequent movement of the water and thus the permeability of the aquifer.

Suspended particles or sand are usually removed by filtering through coarse filter paper before laboratory tests are carried out. Further instructions for the pretreatment of samples can be found in the description of tests for individual parameters in Chapter 6. Each sample should be transferred to several bottles at the place of collection ready for the subsequent laboratory tests to avoid sample division in the laboratory. Each bottle is carefully labelled.

All sampling devices should be completely dried after use where microbiological testing is involved. Growth of algae, bacteria or fungi is thus prevented. Before sampling, the exit point is cleaned or, if possible, flamed with a gas burner. After a running time of at least 5 min a sterile glass bottle of 100 to 1000 ml capacity is filled in free fall. The sampler should position himself downwind of the bottle and avoid coughing or speaking. It is important not to touch the rim or the stopper. Filling is continued until an air bubble of ca. 2 ml remains. This allows the bottle to be shaken more easily later.

To measure temperature, pH, redox potential, conductivity and oxygen directly at the sampling point the appropriate electrodes are immersed in the container into which the water is pumped. To determine the environmental conditions of the water in its original state as precisely as possible (accuracy of the result) a flow-through cell should be used. It is fed by a peristaltic hose pump with a delivery capacity of ca. 100 ml/min, which transports part of the flow from the underwater pump into the cell (Fig. 27). The cell is made of glass (ca. 1 l volume) and has several standard ground glass joints carrying measuring probes. The water flows into the cell through another opening via a two-way glass tube and from there passes into the sample bottle. The oxygen probe is connected before this device if possible, in order to obtain an unaffected oxygen value.

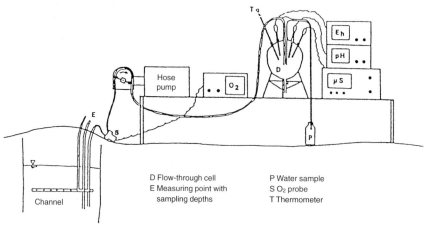

D Flow-through cell   P Water sample
E Measuring point with   S O₂ probe
  sampling depths   T Thermometer

Channel

Fig. 27: Measuring device for the in-situ measurement and sampling of ground water

4.6.2 Surface Water

The collection of surface water samples normally presents no serious technical difficulties. The choice of sampling technique depends on the reason for and the goals of the test. Samples which are taken after pollution incidents or for quality control are usually random samples. For further information on water quality longer series of samples are necessary, which can be taken manually or preferably with automatic samplers. For still bodies of water or slow flowing rivers the collection of depth- or area-integrated samples is a good idea. This involves continuous (pumps) or discontinuous (buckets) removal of individual samples at various depths and locations. Sometimes individual samples are combined to give average samples. The determination of positions of sampling points on the water used to be more expensive than it is now because easy to handle and cost-effective GPS (global positioning satellite) devices are available.

For surface water which is used as crude water for the preparation of drinking water, the frequency of sampling and testing is regulated by the EC guideline 79/869 and supplements to it referring to individual European states. In principle the minimum frequency is greater the larger the quantity of water to be prepared for drinking and the number of inhabitants to be served. The frequency must also increase if the risk to health is increased by a lowering of crude water quality. If smaller quantities of water are used, up to 3 samplings a year are sufficient. For larger quantities and decreased crude water quality, at least 12 are required.

To collect samples, a bucket or sample bottle is simply dipped into the water. In flowing waters the container movement should be against the stream. Here the proximity of still water zones by the bank and stretches of fast running turbulent water often leads to concentration gradients in the content of oxygen and suspended particles. Also a river is not completely mixed below waste water entry points or the joining of tributaries.

During the filling of bottles stoppers or lids should be put down in a clean place. On sampling at bridges attention should be paid to the possibility of whirlpools near the supports affecting the water quality (e.g. oxygen content). Samples for bacteriological examination are collected by dipping the sterile container into the water with the opening against the current. In still water the container is pushed through the water in such a way that the hand has no contact with the water in front of the opening.

4.6.3 Drinking Water

The collection of drinking water samples for physico-chemical examinations is generally problem-free as taps are generally available. Filling several bottles is recommended. For bacteriological tests the tap must be made of metal so that it can be flamed (e.g. with a bunsen burner). Before sampling, water already present in the piping system must be removed by letting the taps run for 15 to 30 min.

Sterile glass bottles of 100 to 1000 ml capacity with standard ground glass joints are used. Contamination of the bottle neck and speaking and coughing during sampling are to be avoided. The bottles are filled until an air bubble of ca. 2 ml remains.

4.6.4 Waste Water

The representative collection of crude water samples with their variable quantities of suspended materials presents problems, especially where automatic devices are employed. Therefore, either exact determination of the solids is dispensed with, or pooled samples are prepared from representative random samples. Where organic substances (e.g. oil) form a separate phase, only manual sample collection can be used.

The sampling of purified waste water is, however, relatively easy and is similar to that of surface water. Random or average samples can be collected (time-, volume- or flow-dependent) either manually or automatically. Modern automatic sampling devices usually permit the programming of different sampling methods and are therefore flexible to use.

Official monitoring of waste water requires a qualified random sample (see Section 4.1) and a two-hourly pooled sample for assessment according to the laws governing water and, in particular, waste disposal. Furthermore, the regulations concerning the location and frequency of waste water sampling depend on the size and technology used in the sewage treatment plant. For manual sampling 15 min pooled samples from several random samples are usually sufficient. For quality assurance two-hourly or daily pooled samples are usually combined. To avoid dividing up samples in the laboratory, several bottles are usually filled if possible.

Highly polluted water leaking from waste dumps is normally collected from waste water shafts or drainage pipes. This water should flow for a longer period through a glass funnel into the overflowing collecting bottle so that the effect of the surrounding air is lowered. If the drainage pipes from a dump open into waste water channels, an automatic sampler can be used to collect time- or flow-dependent individual samples, which can then be pooled.

For representative sampling from waste water pipes with continuous turbulent flow a sampling tube is held in the middle of the main stream. In the case of laminar flow or strongly fluctuating flow rates a perforated tube is placed in the cross section of the pipe and the water is collected manually by opening a valve.

Particular attention should be paid to the hygienic problems involved with waste water sampling and to the safety precautions described in Chapter 1.

4.6.5   Soil

Soil samples can be taken with either disturbed or undisturbed stratification. The former are those which are taken without maintaining the natural structure. Undisturbed samples are removed from the earth carefully using a cutting cylinder so that the structure is retained. The size of a soil sample depends on the test procedure to be used and is generally between 0.3 and 1 kg.

The soil sample must be representative of the whole area to be examined. This requirement is not easy to satisfy even on mixing many individual samples, because spatial heterogeneity can be much more extreme than in the case of water. For collection a random distribution of sampling points would be ideal, but in relatively homogeneous soil pooled samples of almost equally good quality can be obtained with less effort by reducing the sampling area. Suitable sampling methods are illustrated in Fig. 28.

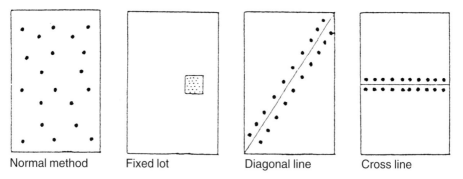

    Normal method   Fixed lot   Diagonal line   Cross line

Fig. 28: Sampling of soils

For the sampling of soil at least 20 to 30 individual samples per hectare are taken using a Pürckhauer drill (or spade if necessary) and combined to give a pooled sample. Atypical features, such as edges of fields, should be avoided. The usual sampling depth is up to 30 cm in arable land or 15 cm in pastures. On sampling from deeper levels it should be ensured that samples are not contaminated by material from upper layers. If undisturbed soil samples are required, a cutting cylinder with a minimum capacity of 100 cm³ is used. For special examinations (e.g. testing of nutrient penetration), samples are taken down to 1 m depth after a pit has been dug out to reveal a smooth, vertical earth profile.

Samples for nutrient testing should be taken where possible at the same time of year, ideally after the harvest and before the application of fertiliser. Samples for testing available nitrogen should only be taken just before fertiliser application in the spring.

For sampling to determine the water profile of soil, a thorough profile description is necessary (e.g. structure, presence of roots, layer changes). The cutting cylinder is pushed steadily in the vertical or horizontal direction to avoid compression.

After collection the samples are packed in plastic bags or bottling jars with lids and seals. All external labelling (place and date, name of sampler, number of sampling point, sampling depth, type of cultivation, weathering) should be stable to environmental effects. For safety reasons each sample should be accompanied by a sheet with the above information.

For longer storage disturbed samples are air dried, milled (stones removed) and sieved. The fine soil (particles < 2 mm diameter) is used in tests.

5 Field Measurements

The quality of test results depends to a large extent on the completeness of the information collected in the field, in addition to sample collecting itself. A check list makes field measurements easier and avoids subsequent time-consuming work. The results of field measurements of rapidly changing parameters are an important part of the necessary information.

5.1 Check List

A check list makes subsequent processing of information easier, particularly when sampling and measuring conditions are difficult. The list shown in Table 17 is not necessarily complete but it acts as a memory aid for planning the work. The data obtained are entered into the sample collection report.

Table 17: Check list for sampling and inspection

Parameter	Ground water	Surface water	Drinking water	Waste water	Soil
Sampling location (O, M)	x	x	x	x	x
Entry of co-ordinates (C)	x	x	x	x	x
Geological conditions (C, O)	x	x			x
Catchment area (O)	(x)	x	x	x	
Surface structure (O)	x	x		x	x
Soil use, vegetation (O)	x	x	x	x	x
Flow velocity (O, M)	(x)	x		x	
Outlet, flow (O, M)		x	x	x	
Sedimentation (O, M)		x		x	
Description of bodies of water (O)					
– Discharge points/effluent inlet points		x		x	
– Organisms	x	x	x	x	
– Eutrophication		x			
– Visible contamination	x	x	x	x	
– Type of spring or well	x	x			
– Signs of corrosion	x	x	x	x	
– Gas evolution	x	x		x	
Soil description (O)					
– Colour					x
– Nature					x
– Type					x
– Compression					x

Table 17: (Continued)

Parameter	Ground water	Surface water	Drinking water	Waste water	Soil
– Root penetration					X
– Humidity					X
Measurements (M, O)					
– Air temperature	X	X		X	
– Air pressure	X	X		X	
– Colour, odour	X	X	X	X	X
– Taste	(X)		(X)		
– Turbidity	X	X	X	X	
– Visible depth		X			
– Sedimentable materials	X	X		X	
– Precipitation	X	X	X	X	
– pH value	X	X	X	X	X
– Redox potential	X	X	X	X	
– Electrical conductivity	X	X	X	X	X
– Oxygen	X	X	X	X	
– Chlorine			X		
– Carbon Dioxide	X		X		
– Aggression	X	(X)	X	(X)	

M = measurement
O = observation
C = chart

5.2   Measurements

5.2.1   Sensory Examination

The sensory examination should take place during sampling because changes in the sample can occur during transport and storage. This test includes odour, taste, transparency, turbidity, and coloration. Soil samples should be tested for odour, colour and consistency on rubbing the moist sample between the fingers.

The *odour* is tested immediately after sampling. Odour strengths and types may be designated as follows:

odour strength: very weak, weak, clear, strong and very strong,

odour type: earthy, mossy, peaty, musty, putrid, reminiscent of manure, fishy, aromatic, or characteristic of a particular substance (e.g. petrol or ammonia).

The qualitative test involves smelling a half-filled, previously shaken bottle.

The *colour* can be tested by viewing the sample in daylight. The designation is as follows: colourless, very weakly coloured, weakly coloured, and strongly coloured. The corresponding colour tone is also given, e.g. yellowish, yellowish-brown, brownish, yellowish-green etc.

The examination of *taste* is only carried out when it is certain that no infectious bacteria or polluting substances are present. Taste sensations can be designated as follows: tasteless, salty, bitter, alkaline, sour, astringent, metallic, repulsive. The degrees of taste may be differentiated by the terms weak, clear and strong.

The *visible depth* of water is defined as the depth at which a white disc (e.g. a Secci disc) let down with a line or pole into the water is just visible. Down to 1 m the values are given in cm intervals, at more than 1 m, 10 cm intervals.

The simple test of *turbidity* involves filling a clean 1 l glass container about two thirds full with the water sample, shaking well and comparing against a black and then a white background. The following degrees of turbidity are determined: clear, opalescent, weakly turbid, strongly turbid, opaque. The measurement of turbidity in the laboratory is described in Section 6.1.34.

5.2.2   Temperature

Areas of Application ➔ air, water, waste water, soil

Apparatus

a) Air Temperature
Calibrated mercury thermometer with 0.5 °C graduations, measurement range –20 to 60 °C
b) Water temperature
Calibrated mercury thermometer with 0.1 °C graduations, measurement range 0 to 100 °C, or electronic thermometer with digital display. A device for showing maximum temperature is useful.
c) Soil Temperature
Where possible a special soil thermometer which is bent at an angle of 30° to facilitate reading; graduation 0.1 °C, measurement range –20 to 60 °C. Alternatively several thermosensors can be used, which only require one common display instrument.

Measurement

The air temperature is measured with a dry thermometer ca. 1 m above the sampling point. The thermometer must be shaded from sunshine.

The water temperature is measured by dipping a mercury thermometer to the reading-off depth and waiting until the reading is constant. Where a direct measurement is not possible (e.g. in wells and springs) a large quantity of water is collected and the measurement taken as quickly as possible. Direct measurements are possible using electronic thermometers with extension cables.

To measure the soil temperature a pointed metal rod with the same diameter as the thermometer is first pushed into the soil. The thermometer is then inserted into the hole down to the desired depth. The thermometer should be protected from the effects of sunshine.

5.2.3   Sedimentable Material

Sedimentable material can be found in some surface water and in untreated waste water. The determination should be carried out in the field immediately after sampling to avoid errors through flocculation. The method is suitable for the determination of sedimentable material above 0.1 ml/l.

Areas of Application → water, waste water

Apparatus
1 l Imhoff sedimentation vessel in glass or transparent plastic
Holding device for the sedimentation vessel

Measurement
1 l of the shaken sample is placed in the sedimentation vessel immediately after collection. After ca. 50 and 110 min sedimentation time the vessel is rotated about the vertical axis, so that any material clinging to the walls sinks to the bottom. After 1 and 2 h the volume of the sediment is read off.

Calculation of Results
The readings are rounded off as shown for the data in Table 18.

Table 18: Rounding off of recorded values for sedimentable materials

Recorded value (ml)	Rounding off of the recorded value (ml/l)
< 2	to 0,1
2–20	to 0,5
10–40	to 1
> 40	to 2

5.2.4   pH Value

The pH value is the negative logarithm to base 10 of the hydrogen ion activity (mol/l) and is 7.0 in pure water (neutral point). This value changes in the presence of acids and alkalis and through the hydrolysis of certain salts. Salts of strong bases and weak acids (e.g. alkali metal carbonates) raise the pH, while salts of weak bases and strong acids (e.g. ammonium chloride) lower it. Acidification of soil can be due to hydrolysis of iron or aluminium salts or by the formation of humic acid during the degradation of organic matter.

In natural water the pH is usually between 6.5 and 8.5. The presence of free carbon dioxide or humus lowers the pH. Biogenic decalcification in surface water, which occurs when there is high CO_2 depletion by algae, can cause the pH to rise to ca. 10.

Areas of Application → water, waste water, soil

Apparatus
a) Indicator papers: for preliminary tests universal indicator paper, 'non-running' pH test strips are better.
b) pH meter with electrode

Reagents and Solutions

Buffer solution pH = 4.62:   200 ml acetic acid, c (CH_3COOH) = 1mol/l, are mixed with 100 ml sodium hydroxide, c (NaOH) = 1mol/l, and 700 ml water.

Buffer solution pH = pH 7.0:   a) 9.078 g potassium dihydrogen phosphate (KH_2PO_4) are dissolved in water and the volume made up to 1 l;
b) 11.88 g disodium hydrogen phosphate ($Na_2HPO_4 \cdot 2H_2O$) are dissolved in water and the volume made up to 1 l; 2 parts a) and 3 parts b) are mixed.

Buffer solution pH = 9.0:   a) 12.40 g boric acid (H_3BO_3) and 100 ml sodium hydroxide, c (NaOH) = 1 mol/l, are dissolved in water and the volume made up to 1 l;
b) hydrochloric acid, c (HCl) = 0.1 mol/l; 8.5 parts a) and 1.5 parts b) are mixed.

Sample Preparation
Water samples require no preparation for pH measurements. For soils the pH of a soil suspension is determined. The soil is shaken with either demineralised water or a calcium chloride solution, c ($CaCl_2$) = 0.01 mol/l, (1 part soil + 2.5 parts liquid) for ca. 30 min. The pH is then measured. Measurements with calcium chloride solution generally give values which are lower than those with demineralised water.

Calibration and Measurement
The pH is measured by dipping the indicator paper or test strips into the solution and comparing the resulting colour with the manufacturer's standards (the development time depends on the manufacturer's instructions).

Before performing potentiometric pH measurements new or dry glass electrodes must be allowed to stand in water or potassium chloride solution, c (KCl) = 3 mol/l, for several days (see manufacturer's instructions.) The calibration is carried out using two standard buffer solutions. The pH of the sample should be between their two pH values. The sample temperature is determined at the same time and is entered into the meter to allow for a temperature correction. Where samples with very different pH values are measured consecutively, the electrode should be placed in water for a long period after the first measurement. The reading is taken after it has remained constant for ca. 1 min. The value can be read accurately to 0.1 units and with sensitive instruments to 0.01 units.

Interfering Factors

At pH values above 10 alkali errors can occur, so the use of alkali-resistant electrodes is recommended. Changes in glass structure can occur in older electrodes resulting in errors, particularly in weakly buffered water.

Oil in the sample can impair the sensitivity. In this case the electrode must be cleaned with soap or surfactants before every measurement. It must then be rinsed with water, dilute hydrochloric acid and then water again.

pH values of lime-containing soils do not always correspond to the pH of the soil solution because the calcium concentration and the partial pressure of the CO_2 are determined with the pH.

Calculation of Results

Below pH 2 and above pH 12 the results are accurate to 0.1 units. Otherwise the results are accurate to a maximum of 2 decimal places, depending on the sensitivity of the instrument.

5.2.5   Redox Potential

To characterise the reducing or oxidising strength of a particular redox couple (e.g. Fe^{2+}/Fe^{3+}) the redox potential is defined as the potential difference between the half element of the redox couple and the standard hydrogen electrode as the arbitrary zero point of the potential scale. To be able to compare the reducing and oxidising strengths of different redox couples directly with one another, a standard state is chosen at which all the redox partners have an activity of a = 1 mol/l at a temperature of 25 °C. This potential is designated as the normal potential.

Like pH values, redox potentials control many chemical processes in water. In bodies of water and waste water anaerobic processes are characterised by low redox potentials. Oxygen enters the surface of still water by diffusion. This creates an aerobic environment in the upper water layer, while as a result of poor mixing there can be an anaerobic environment in the lower layers or in the sediment.

The redox potential measures the competing processes of electron donation (reduction) and acceptance (oxidation).

Areas of Application → water, waste water

Apparatus
pH/mV meter with electrode

Reagents and Solutions
Redox buffer solution: quinhydrone is dissolved in a pH buffer until saturation is reached. The solution must always be freshly prepared. The redox potential is proportional to the pH value in the range 1 to 7, e.g. pH = 4.62, redox potential 427 mV, pH = 7.00, redox potential 285 mV (at 25 °C).

Calibration and Measurement
Control measurements using the quinhydrone buffer solution should be carried out at known time intervals. For measurement, the sample is placed in a container and the electrode set dipped into it.

The reading is recorded after it has remained constant for several min. Changing between solutions with very different ionic activities leads to delays in determining the end point.

Interfering Factors
The recorded value is mainly affected by the ionic activity, temperature and the nature of the electrode surface. When a slow reading is encountered, careful cleaning of the metal ring with talc often helps. Oil contamination is first removed using a household detergent and the electrode is then rinsed with ethanol and water.

Calculation of Results
mV values should only be used to give some idea of the presence of aerobic or anaerobic processes. Below –200 mV processes are strictly anaerobic, between 0 and –200 mV they show transitional character, and positive mV values indicate aerobic processes. A direct comparison of values is only valid for the same redox couples, ionic strengths and pH values. The pH value should always be quoted.

5.2.6 Electrical Conductivity

The electrical conductivity is a total parameter for dissolved, dissociated substances. Its value depends on the concentration and degrees of dissociation of the ions as well as the temperature and the migration velocity of the ions in the electric field.

The results do not give any information on the nature of the ions. However, the concentration of dissolved electrolyte can be calculated from the conductivity when the ionic composition and equivalent conductivities are known.

Electrical conductivity measurements are frequently employed for monitoring surface and ground water and desalination plants. In soil examinations conductivity provides information on the proportion of soluble salts and thus on the suitability of the soil for agricultural use.

The conductivity (unit: $\mu S/cm$) is defined as the reciprocal of the electrical resistance (unit: $S = \Omega^{-1}$) relative to 1 cm^3 water at 25 °C.

Areas of Application → water, waste water, soil

Apparatus
Conductivity meter with electrode, and temperature correction if possible
Thermometer

Reagents and Solutions
Potassium chloride solution: 0.7456 g anhydrous potassium chloride are made up to 1 l with water.

Sample Preparation
No sample preparation is necessary with water samples. For soil samples the measurements are taken on a soil extract, prepared from water and soil in a ratio of 5:1. The soil sample is shaken with the water for at least 2 h in a closed vessel. The sample is allowed to stand for at least 30 min between each shaking process and is finally filtered.

Calibration and Measurement

Before measurement begins the container and cell must be rinsed several times with the solution to be tested. The measurements should be carried out at 25 °C, otherwise a temperature correction must be calculated. (Correction factors are given in the manufacturer's instructions or in Table 22, Section 6.1.6).

To check the instrument the cell constant is measured from time to time using standard potassium chloride solutions. The manufacturer's instructions should also be followed for this procedure.

After measuring oil-containing samples the electrode should be cleaned thoroughly with a solvent (e.g. acetone).

Calculation of Results

The ionic strength I is calculated from the electrical conductivity using the following equation:

$$I = 1.83 \cdot \chi_{20} \cdot 10^{-5}$$

χ_{20} is the electrical conductivity at 20 °C in $\mu S/cm$.

This formula is only valid for water containing carbonate. For soil the calculation of salt concentrations is only an approximation because of the variety of salt compositions.

In 1 l of a water sample or soil extract the concentrations of some pure salts (without water of crystallisation) at 1000 $\mu S/cm$ are as follows:

magnesium chloride ($MgCl_2$)	0.40 g
calcium chloride ($CaCl_2$)	0.44 g
sodium chloride ($NaCl$)	0.51 g
sodium sulfate (Na_2SO_4)	0.62 g
magnesium sulfate ($MgSO_4$)	0.72 g
calcium sulfate ($CaSO_4$)	0.80g.

For many soils a conductivity of 1000 $\mu S/cm$ of an extract corresponds to ca. 65 mg salt in 100 ml solution. The salt content of the dry soil sample can be calculated taking into account the quantity of water used for elution and the mixing ratio.

5.2.7   Oxygen

Oxygen is essential for the survival of most organisms in water. This also applies to the metabolic pathways of aerobic bacteria and other micro-organisms, which are responsible for the degradation of pollutants in water and utilise oxygen as an electron acceptor for this purpose.

Oxygen reaches the water via surface diffusion and by photosynthesis in algae and submerged plants. When plant growth is prolific, as with the recent algae blooms, supersaturation by oxygen can occur. In drinking water an oxygen concentration of at least 4 mg/l is necessary to prevent corrosion of the carrier pipes.

Oxygen can be determined amperometrically or titrimetrically using the modified Winkler method.

Areas of Application ➔ water, waste water

Apparatus
a) Amperometric Determination
Oxygen-measuring instrument with electrode
b) Winkler Method (Titrimetric)
Ground glass bottles of known capacity (110 to 150 ml)
Glass apparatus for volumetric anaysis

Reagents and Solutions
a) Amperometric Determination

Control solution:	freshly prepared saturated sodium dithionite solution ($Na_2S_2O_4$)
Air-saturated solution:	demineralised water is aerated for a prolonged period.

b) Winkler Method

Manganese(II) sulfate solution:	480 g manganese(II) sulfate ($MnSO_4 \cdot 4H_2O$) (or 400 g $MnSO_4 \cdot 2H_2O$) are made up to 1 l with water.
Alkali iodide – azide solution:	350 g sodium hydroxide (or 500 g potassium hydroxide), 150 g potassium iodide (or 135 g sodium iodide) and 1 g sodium azide are made up to 1 l with water.
Sodium thiosulfate solution, $c\,(Na_2S_2O_3) = 0.01\,mol/l$:	freshly prepared by diluting sodium thiosulfate solution, $c\,(Na_2S_2O_3) = 0.1\ mol/l.$
Orthophosphoric acid:	at least $w\,(H_3PO_4) = 85\%$
Starch solution:	1 g soluble starch is boiled and a few drops of formalin are added.

Calibration and Measurement
a) Amperometric Determination
The oxygen electrode is first placed in the control solution until a constant reading is obtained. The electrode is then quickly washed and placed in the air-saturated solution until the reading is constant. Further adjustment is based on the air pressure and water temperature according to Table 19. The electrode is then ready. The calibration can be carried out less accurately by placing in the control solution followed by exposure of the dry electrode to air. The instrument is adjusted according to the partial pressure of the oxygen in the air (see manufacturer's instructions).

Table 19: Oxygen saturation of water as a function of temperature and air pressure

Water temperature (°C)	Oxygen saturation (mg/l) at air pressure (hPa $\cong$ mbar)				
	933	960	986	1013	1040
0	13.41	13.80	14.18	14.57	14.95
2	12.70	13.06	13.43	13.79	14.16
4	12.04	12.38	12.73	13.08	13.42
6	11.43	11.76	12.09	12.42	12.75
8	10.87	11.19	11.50	11.81	12.13
10	10.36	10.66	10.96	11.26	11.56
12	9.88	10.17	10.46	10.74	11.03
14	9.45	9.72	9.99	10.27	10.54
16	9.04	9.31	9.57	9.83	10.10
18	8.67	8.92	9.18	9.43	9.68
20	8.33	8.57	8.81	9.06	9.30
22	8.01	8.24	8.48	8.71	8.95
24	7.71	7.94	8.16	8.39	8.62
26	7.43	7.65	7.87	8.09	8.31
28	7.17	7.38	7.60	7.81	8.02
30	6.93	7.13	7.34	7.55	7.76
32	6.70	6.90	7.10	7.30	7.50
34	6.48	6.67	6.87	7.07	7.26
36	6.27	6.46	6.65	6.84	7.03

b) Winkler Method
In this method the container is filled using a tube free of air bubbles and rinsed several times to overflowing. For oxygen fixation the following are pipetted successively under the sample surface in a full container: 0.1 ml manganese sulfate solution and 0.5 ml iodide-azide solution. The container is closed free of air bubbles and shaken. In the laboratory 2 ml phosphoric acid are also pipetted in, the container closed and shaken again. After ca. 10 min the contents of the container are transferred to an Erlenmeyer flask and titrated with sodium thiosulfate solution, c ($Na_2S_2O_3$) = 0.01 mol/l. When only a weak yellow coloration remains, 1 ml of the starch solution is added and titration continued until the blue colour disappears.

Interfering Factors
When the control values or saturation point are found to lie outside the normal measuring range, the measuring head with the inner electrode, working electrode and membrane must be cleaned or replaced. The main interfering substance is hydrogen sulfide. In the Winkler method interference by iron(III) and nitrite ions is inhibited by the addition of phosphoric acid and azide during the determination.

Calculation of Results

The concentration of oxygen dissolved in the water sample is calculated from the following equation:

$$\beta(O_2) = V_T \cdot c \cdot f/V_P \cdot V_F/(V_F - V_R)$$

V_T volume of sodium thiosulfate solution consumed in the titration, ml
c concentration of the sodium thiosulfate solution, mol/l, (here: c ($Na_2S_2O_3$) = 10 mmol/l)
f equivalence factor with the unit mg/mmol (here : f = 8 mg/mmol)
V_P volume of analysis sample used in the titration, ml
V_F volume of sample container, ml (here: $V_F = V_P$)
V_R total volume of reagent solutions added, ml (here: V_R = 0.6 ml)

Oxygen saturation (%):

$$\beta(O_2) = a \cdot 100/b$$

a measured oxygen concentration, mg/l
b oxygen saturation concentration at the temperature measured, mg/l

5.2.8 Chlorine

Chlorine is used as a disinfectant for the treatment of drinking water, bathing water and, in special cases, also for waste water. Chlorine in the form of dissolved elemental chlorine, hypochlorous acid and hypochlorite ions is known as free chlorine. Chlorine compounds which are formed by reaction of hypochlorite ions with ammonium or organic compounds containing amino groups are known as bound chlorine. Both together are known as active chlorine, free chlorine being the stronger oxidising agent.

Active chlorine should be determined at each stage of drinking water treatment, both in the mains and at the point of consumption to guarantee bacteriologically impeccable water. 0.2 to 0.5 mg/l active chlorine should be present in drinking water.

A field method using a comparator and the titrimetric DPD method (N,N-diethyl-p-phenylene-diamine, $C_{10}H_{16}N$) are described below.

Areas of Application → water, waste water

Apparatus
a) Field Method
Comparator (e.g. Lovibond-Tintometer®)
Comparator filters β (Cl_2) = 0.1 to 1 mg/l and 1 to 4 mg/l
DPD tablets nos. 1 and 3 (e.g. supplier: Lovibond-Tintometer, UK)
b) Titration Method
Glass equipment for volumetric analysis

Reagents and Solutions

Glycine solution:	20 g glycine ($C_2H_5NO_2$) are dissolved in 200 ml water.
Buffer solution:	24 g disodium hydrogen phosphate (Na_2HPO_4) and 46 g potassium dihydrogen phosphate (KH_2PO_4) are dissolved in ca. 800 ml water; 100 ml 0.8 % EDTA solution ($C_{10}H_{14}N_2O_8Na_2 \cdot 2H_2O$) are added and the volume made up to 1 l.
DPD solution:	1.5 g DPD sulfate are dissolved in ca. 800 ml water; 8 ml 40 % sulfuric acid and 25 ml 0.8 % EDTA solution are added and the volume made up to 1 l; the solution is stored in brown bottles; if coloration appears it is no longer usable.
FAS solution:	1.106 g ferrous ammonium sulfate (($NH_4)_2Fe(SO_4)_2 \cdot 6H_2O$) are dissolved in 800 ml water; 1 ml 40 % sulfuric acid is added and the volume made up to 1 l. The solution can be kept for ca. 1 month. Titration is carried out according to Section 6.1.7b.

Potassium iodide, solid

Calibration and Measurement

a) Field Method with Comparator

The measuring cuvette (10ml) is washed with the sample water and the no. 1 tablet is placed in it and then dissolved in a little sample water. The solution is made up to 10 ml, mixed and the cuvette placed in the comparator together with the comparison cuvette containing pure water. The apparatus is held against white light and adjusted by means of the thumb wheel until the colours are identical. The reading is then noted.

Tablet no. 3 is placed in the sample cuvette, mixed, and left for 2 min. The new value giving the same colour is noted. The first value gives the free chlorine and the difference between the values the bound chlorine.

b) Titration Method

Titration 1 – Determination of Total Chlorine

5 ml glycine solution and 200 ml water sample are placed in an Erlenmeyer flask and left for 2 min. The solution is then poured into a second Erlenmeyer flask containing 10 ml buffer solution and 5 ml DPD solution. The contents are then titrated with FAS solution until colourless.

Titration 2 – Determination of Free Chlorine

10 ml buffer solution, 5 ml DPD solution and 200 ml water sample are placed in an Erlenmeyer flask. After 5 min the contents are titrated with FAS solution until colourless.

Titration 3 – Determination of Bound Chlorine

The solution from titration 2 is treated with 1 g solid potassium iodide. After dissolving and leaving for 5 min, the solution is titrated with FAS solution until colourless.

Interfering Factors

Large amounts of manganese dioxide cause interference. Higher concentrations of copper and iron ions form complexes with the EDTA solution used.

Calculation of Results

Free and bound chlorine (mg/l) are calculated according to:

β (free chlorine) = (consumption from titration 2/consumption from titration 1) $\cdot$ 0.5
β (bound chlorine) = consumption from titration 3 $\cdot$ 0.5.

If chlorine dioxide (ClO_2) is used as the chlorinating agent in drinking water, the following equation applies:

β (chlorine dioxide) = consumption from titration 1 $\cdot$ 0.95

5.2.9   Acidic Capacity

The acidic capacity (K_A in mmol/l) is defined as the capacity of substances contained in the water to take up hydroxonium ions (H_3O^+) to reach a defined pH value. In natural water the hydroxonium ions are bound mainly by anions of weak acids (mainly carbonate and hydrogen carbonate). At high pH values in chalk-rich water the suspended calcium carbonate can be determined by the acidic capacity.

Two parameters can be distinguished:

– acidic capacity to pH 8.2 $\rightarrow K_{A\,8.2}$
– acidic capacity to pH 4.3 $\rightarrow K_{A\,4.3}$.

$K_{A\,8.2}$ is determined in water with pH above 8.2 and $K_{A\,4.3}$ in water with pH above 4.3. Depending on the accuracy requirements, the measurement can be made potentiometrically or using colour indicators.

Areas of Application ➔ water, waste water

Apparatus
pH meter with electrode
Magnetic stirrer
Titration equipment

Reagents and Solutions

Phenolphthalein indicator solution:	1 g phenolphthalein is dissolved in 100 ml ethanol-water mixture (1 + 1).
Mixed indicator:	0.02 g methyl red and 0.1 g bromocresol green are dissolved in 100 ml ethanol.

Hydrochloric acid, c (HCl) = 0.1 mol/l
$\qquad\qquad\qquad c$ (HCl) = 0.02 mol/l

Sample Preparation
The end point determination with colour indicators can be hindered by coloration and turbidity. Free chlorine, chlorine dioxide and ozone can destroy the indicator. Contaminants causing turbidity are removed by filtration (membrane filter, pore size 0.45 µm), while coloration can be reduced by the addition of activated charcoal followed by filtration. Free chlorine is reduced by the addition of one drop of sodium thiosulfate solution, c ($Na_2S_2O_3$) = 0.1 mol/l.

Measurement
100 ml of the water sample is placed in an Erlenmeyer flask. With potentiometric measurement the pH electrode is placed in the water and titration with hydrochloric acid, c (HCl) = 0.1 mol/l, is carried out until the pH value has reached 8.2 (for starting values > 8.2) and the mixture is then left for 2 min. The volume of acid consumed is noted and the titration continued to pH 4.3. The solution is stirred or shaken during the whole titration. If the acid consumption is small, the titration is repeated using hydrochloric acid, c (HCl) = 0.02 mol/l.

For measurements involving a colour indicator 2 to 3 drops of phenolphthalein solution are first added. On obtaining a red colour, the sample is first titrated until colourless; 2 to 3 drops of the mixed indicator are then added and titration is continued until the solution turns from green to red.

Interfering Factors
Interference through coloration, turbidity or free chlorine has already been mentioned. In addition absorption or loss of carbon dioxide during or after sampling can falsify the results. Dissolved silicates, phosphates, borates or humic acid salts are also determined by these measurements. Their presence is not an interference but they must be taken into account during the calculation of the concentrations of carbon dioxide, hydroxide or carbonate.

Calculation of Results
$K_{A8.2}$ and $K_{A4.3}$, given in mmol/l, are calculated as follows:
$$K_{A8.2} = V_1 \cdot c \cdot 1000/s$$
$$K_{A4.3} = V_2 \cdot c \cdot 1000/s$$

Where V_1 and V_2 are the volumes of acid consumed in ml to pH 8.2 and 4.3 respectively, c is the concentration of the acid in mol/l and s is the volume of the sample in ml.

For a sample volume of 100 ml and a hydrochloric acid concentration, c (HCl) = 0.1 mol/l, the acid consumption gives the acidic capacity $K_{A8.2}$ or $K_{A4.3}$ directly.

5.2.10   Base Capacity

The base capacity (K_B, in mmol/l) is defined as the capacity of substances contained in water to take up hydroxide ions (OH^-) to reach a defined pH value. In natural water this occurs mainly through carbonic acid (sometimes also humic acids) so that the base capacity mainly shows the concentration of dissolved carbon dioxide in water.

Two parameters can be distinguished:

– base capacity up to pH 4.3 → $K_{B\,4.3}$
– base capacity up to pH 8.2 → $K_{B\,8.2}$

$K_{B\,8.2}$ is determined far more frequently, and $K_{B\,4.3}$ only in special cases (peat water, open drain water and water from cation exchangers). Depending on the accuracy requirements, the measurement can be carried out potentiometrically or using colour indicators.

Areas of Application → water, waste water

Apparatus
pH meter with electrode
Magnetic stirrer
Titration equipment

Reagents and Solutions

Phenolphthalein indicator solution:	1 g phenolphthalein is dissolved in 100 ml ethanol-water mixture (1 + 1).
Methyl orange solution:	0.05 g methyl orange are dissolved in 100 ml water.
Potassium sodium tartrate solution:	50 g potassium sodium tartrate ($C_4H_4O_6KNa \cdot 4H_2O$) are made up to 100 ml with water. After 1 to 2 days the pH is adjusted to 8.2.

Sodium hydroxide, c (NaOH) = 0.1 mol/l
$\qquad\qquad\quad c$ (NaOH) = 0.02 mol/l

Measurement
100 ml of the water sample are placed carefully in an Erlenmeyer flask by allowing the contents of the measuring cylinder or pipette to run slowly down the wall. For potentiometric titrations the pH electrode is placed in the flask and titration is carried out with sodium hydroxide, c (NaOH) = 0.1 mol/l, with careful stirring or swirling until pH 4.3 is reached (if the pH of the sample is below 4.3). If the pH is above 4.3, ca. 1 ml potassium sodium tartrate solution is added and the solution is titrated to pH 8.2. The pH values should remain constant for ca. 2 min. The volume of alkali consumption is noted and the process is repeated with sodium hydroxide, c (NaOH) = 0.1 mol/l or for low consumption with c (NaOH) = 0.02 mol/l, whereby the total amount of solution consumed in the first determination is added at once.

In the titration using colour indicators $K_{B\,4.3}$ is titrated with 2 to 3 drops of methyl orange until the colour changes from orange to yellow. Separately from this, $K_{B\,8.2}$ is titrated after addition of 5 drops of phenolphthalein solution until the colour changes from colourless to pink.

Water samples with a high carbon dioxide content are placed directly in graduated titration vessels (if necessary graduated beakers or measuring cylinders). A defined volume of sodium hydroxide, c (NaOH) = 0.1 mol/l, together with 1 ml potassium sodium tartrate solution are added and the resulting solution back-titrated with hydrochloric acid, c (HCl) = 0.1 mol/l.

Interfering Factors
Interference caused by oxidisable or hydrolysable ions of iron, aluminium or manganese is inhibited by the potassium sodium tartrate solution. Loss of carbon dioxide during sample collection and titration can lower the value of $K_{B\,8.2}$.

Coloration and turbidity of the water sample, as well as free chlorine can cause interference. Pretreatment as described in Section 5.2.9 (Acidic Capacity) can be carried out. Loss of carbon dioxide during sampling should be avoided, e.g. by immersing the pipettes.

Calculation of Results
$K_{B\,4.3}$ and $K_{B\,8.2}$ are given in mol/l and are calculated as follows:

$$K_{B\,8.2} = V_1 \cdot c \cdot 1000/s$$
$$K_{B\,4.3} = V_2 \cdot c \cdot 1000/s$$

V_1 and V_2 are the volumes in ml of the sodium hydroxide consumed up to pH 4.3 and 8.2 respectively, c the concentration of sodium hydroxide in mol/l, and s the volume of the sample in ml.

With a sample volume of 100 ml and the use of sodium hydroxide, c (NaOH) = 0.1 mol/l, the numerical value of the consumption of alkali gives the base capacity $K_{B\,4.3}$ or $K_{B\,8.2}$ directly.

5.2.11   Calcium Carbonate Aggression

The rapid test described below has been developed for the preliminary estimation of the calcium aggressive or calcium carbonate precipitative properties of crude water used for the preparation of drinking water. It is based on the reaction between dissolved carbon dioxide and solid calcium carbonate.

The equilibrium between the solid calcium carbonate and dissolved carbon dioxide is described by the following equations:

$$H_2CO_3 + H_2O \rightleftharpoons H_3O^+ + HCO_3^-$$
$$HCO_3^- + H_2O \rightleftharpoons H_3O^+ + CO_3^{2-}$$
$$CaCO_3 \rightleftharpoons Ca^{2+} + CO_3^{2-}$$

The solution has a particular pH as a consequence of these reactions. If solid calcium carbonate is added to water which is not in equilibrium the following can occur:

- The pH value increases; i.e. the water is chalk aggressive.
- The pH value decreases, i.e. the water is chalk precipitative.

For many purposes the following rapid test is adequate for the assessment of water. For more exact measurements (e.g. for planning deacidification plants) other methods are necessary (see Section 6.1.6 'Calcium Carbonate Saturation and Equilibrium pH').

Areas of Application ➜ water

Apparatus
pH meter with electrode
Tapered conical flask made of acrylic glass, ca. 50 to 100 ml capacity (Fig. 29) (if not available a beaker may be used)
Thermometer

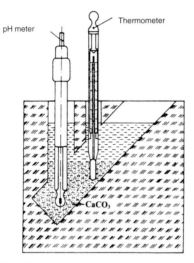

Fig. 29: Vessel for the determination of calcium carbonate aggression

Reagents and Solutions
Marble powder, w ($CaCO_3$) = 99%
Hydrochloric acid, w (HCl) = 10%

Measurement
The pH electrode is dipped into the tip of the conical container. Sample water is passed through the container until a constant pH is obtained. Marble powder is added until the tip of the electrode is completely covered. After ca. 2 min the pH is read again. The temperature of the marble powder should be the same as that of the water. After each measurement the container is cleaned with dilute hydrochloric acid.

Calculation of Results
The difference between the two measured pH values is calculated. Water with a Δ pH of up to $\pm$ 0.04 is considered to have a chalk – carbon dioxide equilibrium (Table 20).

Table 20: Classification of calcium carbonate aggression in water

pH increase	Aggression	pH decrease	Precipitation
+ 0.04 up to + 0.1	Essentially equilibrium	– 0.04 up to – 0.1	Essentially equilibrium
+ 0.1 up to + 0.5	Weakly aggressive	– 0.1 up to – 0.5	Calcium carbonate precipitation possible
+ 0.5 up to + 1	Aggressive	– 0.5 up to – 1	Calcium carbonate precipitation
> + 1	Strongly aggressive	< – 1	Extensive calcium carbonate precipitation

6 Laboratory Measurements

Water and waste water samples generally require no special sample preparation. Care should be taken during sample collection and processing of water because its contents can easily change (see Chapter 4). It is generally desirable to carry out analyses as soon as possible after sample collection, in order to avoid falsification of results.

In some samples heavy metals can be complexed to humic substances in varying degrees. In these cases it is recommended that nitric acid or a nitric acid – hydrochloric acid mixture (3:1) is added to the sample. It should then be carefully evaporated to a smaller volume to break down these complexes. This also allows the detection limit to be lowered, provided that matrix problems related to increased salt concentration do not arise. Metals adsorbed on to suspended particles are usually oxidised and brought into solution by this type of pretreatment.

If the determination of the chemical or biochemical oxygen demand of waste water includes that of suspended solids, the sample must be homogenised, e.g. with a high-speed stirrer.

The final analysis result depends greatly on sample preparation and processing. It should therefore always be stated whether the sample was tested in the original state or homogenised, sedimented, centrifuged or filtered. For example, rapid shaking of a waste water sample cannot guarantee the uniform distribution of sedimentable and suspended material. It is preferable to use a magnetic stirrer or best of all a high-speed stirrer for homogenisation.

Any necessary pretreatment steps are described together with individual analytical methods.

6.1 Chemical and Physical Analytical Methods

6.1.1 Adsorbable Organic Halogen Compounds (AOX)

Of all the organohalogen compounds, the chlorinated hydrocarbons are the most important. Concentrations in the µg range can be found in surface water and some drinking waters. In waste water, particularly that of industrial origin, concentrations of > 1 mg/l are not uncommon. Certain organohalogen compounds are discharged directly into bodies of water or waste water, e.g. pesticide residues from agriculture or components of commercial or household cleaning agents. In addition, under unfavourable conditions organochlorine compounds can be formed during chlorination of drinking water (here mainly volatile substances in the presence of humic substances) or during the disposal of chlorine-releasing cleaning agents and disinfectants into waste water.

For routine tests or within the framework of official monitoring processes, the total parameter AOX is frequently determined. The analytical method is based on the adsorption of organohalogen compounds on activated charcoal. This process is, however, not always complete and depends on particular molecular properties, such as polarity, molecular mass and the nature of the functional groups. Volatile halogenated solvents are only partially determined and should therefore be analysed by gas chromatography or, after degassing, with appropriate gas detection tubes (the latter is only a screening process).

The following description of the shaking method is based on DIN 38409, part 14 (DIN EN 1485).

Instead of a microcoulometric or ion chromatography determination of the chloride, classical methods of determination can also be used, some of which are less precise.

Areas of Application → water, waste water

Apparatus
Combustion apparatus, suitable for AOX determination
Instrument for microcoulometric chloride determination
Device for membrane filtration, polycarbonate 0.45 μm filter
pH meter with electrode
Shaker

Reagents and Solutions

Activated charcoal:	specially designed for AOX determinations
Nitrate stock solution:	17 g sodium nitrate are dissolved in water and the solution made up to 100 ml with water after addition of 2 ml conc. nitric acid.
Nitrate wash solution:	50 ml of the stock solution are made up to 1 l with water.

Chloride standard, $\beta\,(Cl^-) = 1$ mg/l

Calibration and Measurement
5 ml stock nitrate solution are added to 100 ml of the water sample in a 250 ml Erlenmeyer flask and the pH adjusted to 2 to 3 with nitric acid. The concentration of organic substances in the sample should be below a DOC content of 10 mg/l (≈ a COD of ca. 25 mg/l) and the chloride concentration below 1000 mg/l. The mixture should be diluted if necessary. 50 mg activated charcoal are added, the mixture shaken for at least 1 h in a shaker and the suspension filtered through a membrane filter. The filter residue is washed several times with 50 ml nitrate solution. The moist residue and filter are placed together in a quartz boat and the latter carefully pushed into the tube of a combustion apparatus. The combustion is carried out at 950 °C according to the manufacturer's instructions. The combustion gases are collected in an absorption vessel filled with dilute sulfuric acid. The chloride concentration is usually determined coulometrically or by using an ion chromatograph. A blank consisting of 50 mg unloaded activated charcoal washed with nitrate solution is analysed in the same way.

Interfering factors
Chlorine-containing suspended particles in the sample can lead to values which are too high. The same applies to dissolved bromide or iodide. A proportion of volatile components can escape during sampling, transport or homogenisation, leading to values which are too low. Interference can occur if the DOC is > 10 mg/l or Cl^- > 1000 mg/l, so in these cases test solutions must be diluted, sometimes resulting in the lower detection limit of 10 μg/l not being reached.

Calculation of Results
The content of adsorbable halogen compound in the sample is determined as chloride and given as AOX in μg/l.

6.1.2   Ammonium

Ammonium ions can be formed in water and soil both by microbiological degradation of nitrogen-containing organic compounds and by nitrate reduction under defined conditions. Considerable concentrations of up to 50 mg/l are present in household waste water and very high concentrations of up to 1000 mg/l in water seeping from waste dumps. For this reason ammonium is regarded, with some limitations, as a contamination indicator in ground or drinking water.

If ammonium-containing water is brought into contact with oxygen over a prolonged period the ammonium can be oxidised microbiologically to nitrate via nitrite.

In aqueous solutions there is a pH-dependent equilibrium between free ammonia (toxic to fish) and ammonium ions (Fig. 30).

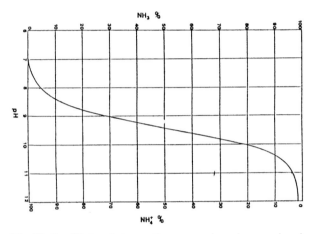

Fig. 30: Equilibrium between free ammonia and ammonium ions as a function of pH value

Areas of Application → water, waste water, soil

Apparatus
Spectrophotometer or filter photometer with 655 nm filter
Water bath

Reagents and Solutions

Salicylate-citrate solution: 32.5 g Na salicylate ($C_7H_5O_3Na$) and 32.5 g trisodium citrate ($C_6H_5O_7Na_3 \cdot 2H_2O$) are dissolved in ca. 200 ml water. 0.243 g disodium pentacyanonitrosylferrate ($Na_2Fe(CN)_5NO \cdot 2H_2O$) are added and the volume made up to 250 ml. The solution can be kept in the dark for ca. 2 weeks.

| Reagent solution: | 3.2 g sodium hydroxide are dissolved in 50 ml water. After cooling, 0.2 g sodium dichloroisocyanurate ($C_3N_3Cl_2ONa$) are added and the solution made up to 100 ml. The solution must be freshly prepared daily. |
| Standard ammonium solution $\beta(NH_4^+) = 1$ g/l: | 2.966 g ammonium chloride (dried at 105 °C) are made up to 1 l with water. A series of diluted solutions for the calibration curve are prepared from this solution. |

Sample Preparation

Samples with suspended particles are filtered (0.45 µm membrane filter). With coloured water or waste water flocculation with aluminium salts can help.

Calibration and Measurement

Depending on the expected ammonium content, up to 40 ml of the sample are pipetted into a 50 ml volumetric flask. 4 ml salicylate-citrate solution are added and the mixture shaken. The pH should be 12.6, which is usually the case with measurements with neutral water samples. 4 ml reagent solution are then added, and the flask filled, shaken and placed in a water bath at 25 °C. After 1 h the solution is analysed at a wavelength of 655 nm in the photometer.

A calibration curve is drawn using the stock ammonium chloride solution in the range 5 to 50 µg/(40 ml of sample). A blank is treated in the same way as the sample.

Interfering Factors

The inorganic substances contained in the water do not cause interference at normal concentrations, nor does urea. However, amines can cause interference in smaller concentrations.

Calculation of Results

The results are calculated from the calibration curve.

6.1.3   Biochemical Oxygen Demand

The biochemical oxygen demand (BOD) is defined as the quantity of dissolved oxygen which is able to oxidise the organic components in the water with the assistance of micro-organisms under defined experimental conditions. The BOD is an empirical biological test in which the water conditions, such as temperature, oxygen concentration or type of bacteria, play a decisive role. These and other factors therefore cause the reproducibility to be much less than that of purely chemical tests. In spite of this disadvantage, the BOD is of considerable importance in the assessment of polluted surface water and waste water. Knowledge of the BOD is indispensable for the planning and building of sewage works.

Biochemical degradation in waste water frequently takes place in two virtually distinct phases. In the first phase organic compounds are decomposed and in the second, known as the nitrification phase, ammonium is oxidised to nitrite and then nitrate by the action of the bacteria species *Nitrosomas* and *Nitrobacter* respectively. This nitrification does not take place in every water or waste water sample, so the test results can be uncertain. Nitrification is frequently observed in treated waste water from sewage works since here nitrifying bacteria will have already multiplied. To achieve better comparability between series of measurements, nitrification can be prevented by the action of inhibitors. However, these should not be used in tests on surface water, because the total oxygen depletion, not just that caused by organic substances, is usually being assessed. For assessment of BOD in connection with water quality, see Section 3.2.

Normally a reaction time of 5 days is used for the measurement (BOD_5). The dilution method is described below, in which the sample is treated with oxygen-saturated dilution water. Manometric measuring systems are commercially available and give useful results. However, the data from the two techniques are not directly comparable. It is also not permissible to convert results obtained for a particular incubation period (e.g. BOD after 5 days) into results for other lengths of time except with considerable limitations.

Besides the procedures referred to, cuvette tests for BOD determination are also commercially available and are particularly suitable for plant laboratories in sewage works.

Areas of Application → water, waste water

Apparatus
250 ml narrow-necked glass bottles with ground glass stoppers
Equipment for maintaining 20 °C (water bath, incubator)
Oxygen-measuring instrument with electrode
pH meter with electrode

Reagents and Solutions

Nutrient salt solutions:	a) 42.5 g potassium dihydrogen phosphate (KH_2PO_4) are dissolved in ca. 700 ml water and 8.8 g sodium hydroxide are added, followed by 2 g ammonium sulfate. The solution is made up to 1 l and the pH adjusted to 7.2.
	b) 22.5 g magnesium sulfate ($MgSO_4 \cdot 7H_2O$) are dissolved in water and the solution made up to 1 l.
	c) 27.5 g calcium chloride ($CaCl_2 \cdot 6H_2O$) are dissolved in water and the solution made up to 1 l.
	d) 0.15 g iron(III) chloride ($FeCl_3 \cdot 6H_2O$) are dissolved in water and the solution made up to 1 l.
Dilution water:	demineralised water is employed. 1 ml of each of the nutrient solutions a) to d) is added to 1 l water and aeration carried out for several days in the dark.
N-allylthiourea solution:	ca. 1 mg N-allylthiourea ($C_4H_8N_2S$) is made up to 100 ml with water. The solution needs to be freshly prepared each day.

Sample Preparation
The pH value of the sample is adjusted to 7 or 8 with hydrochloric acid or sodium hydroxide where necessary. Solids can be included in the determination. However, samples are usually tested after 2 h sedimentation or after filtration. Cooled samples are warmed to room temperature before testing.

Measurement

The water sample is left undiluted up to an expected BOD_5 of $\beta(O_2) = 6$ mg/l. In the case of water with a low bacterial count 5 ml of run-off from a biological sewage treatment plant or, instead, 1 ml of sedimented crude waste water is added to 1 l of the dilution water. The dilution should be carried out in such a way that after 5 days incubation time at least 2 mg/l oxygen have been consumed while the remaining oxygen concentration does not fall below 2 mg/l. Since the final BOD is not known, several dilutions should be prepared so that at least one falls in a favourable measuring range. A previously determined COD value can be useful in selecting the dilution, by dividing this value by 2 and selecting the dilution from Table 21.

After dilution the sample is mixed and carefully transferred to the test bottles, avoiding air bubble formation. The bottles are left for a short time and any air bubbles are removed by tapping the bottle carefully. The ground glass stoppers are then inserted without creating any further air bubbles.

Table 21: Dilution series in the determination of biochemical oxygen demand

Expected BOD_5, $\beta(O_2)$ (mg/l)	Quantity of sample (ml) made up to 1 l
up to 6	1000
4–12	500
10–30	200
20–60	100
40–120	50
100–300	20
200–600	10
400–1200	5
1000–3000	2
2000–6000	1

The oxygen concentration is determined immediately on one of the triplicate (minimum) samples using the oxygen electrode or the Winkler method (see Section 5.2.7). The remaining bottles are kept for 5 days in the dark at 20 °C. The residual oxygen concentration is then measured. A blank, consisting of the inoculated dilution water, is measured in parallel.

Interfering Factors

Any undesired oxygen consumption caused by nitrification can be inhibited by adding 1 ml of an N-allylthiourea solution. Free chlorine, present in some waste water after chlorination, reacts with organic components within about 2 h and then does not interfere further. Chemical reducing agents (e.g. iron(II), sulfite or sulfide ions) are oxidised by leaving the original sample for 2 h with occasional shaking. Substances in the waste water which are toxic to bacteria can inhibit degradation and result in low BOD values, despite the presence of sufficient degradable organic matter. In these cases the samples should be more highly diluted.

Calculation of Results

The biochemical oxygen demand, expressed as $\beta\,(O_2)$ is given in mg/l and calculated according to:

$$\beta\,(O_2) = A/B \cdot (C\!-\!D) + D$$

A   total volume after dilution, ml
B   volume of undiluted sample, ml
C   oxygen consumption of diluted sample after 5 days, mg/l
D   oxygen consumption of dilution water after 5 days, mg/l

6.1.4   Boron

Boron occurs in natural, unaffected water, usually only at very low concentrations, while in domestic waste water concentrations of several mg/l are not uncommon owing to the perborate content of detergents. Such small concentrations are not harmful to human beings but can damage certain plants, e.g. citrus fruits or beans, when present in irrigation water.

The analysis method using azomethine-H is described below.

Areas of Application ➔ water, waste water, soil

Apparatus
Spectrophotometer or fixed filter photometer with 414 nm filter

Reagents and Solutions

Azomethine-H solution:	1 g azomethine-H sodium salt ($C_{17}H_{12}NNaO_8S_2$) and 3 g L-(+)-ascorbic acid ($C_6H_8O_6$) are made up to 100 ml with water. The solution is stored in a plastic bottle and is stable for ca. 1 week in the refrigerator.
Buffer solution, pH = 5.9:	25 g ammonium acetate, 25 ml water, 8 ml sulfuric acid, $w\,(H_2SO_4) = 29\%$, 0.5 ml phosphoric acid, $w\,(H_3PO_4) = 85\%$, 100 mg citric acid ($C_6H_8O_7 \cdot H_2O$) and 100 mg disodium EDTA ($C_{10}H_{14}N_2Na_2O_8 \cdot 2H_2O$) are mixed.
Reagent solution:	equal volumes of azomethine-H solution and buffer solution, pH = 5.9, are mixed before the analysis. The solution is stored in a refrigerator.
Standard borate solution $\beta\,(BO_3{}^{3-}) = 1$ mg/l:	572 mg boric acid (H_3BO_3) are made up to 1 l with water. 10 ml of this solution are taken and made up to 1 l with water.

Sample Preparation
Suspended particles should be removed by filtration through a 0.45 µm membrane filter.

Calibration and Measurement

0.5 to 6 ml of the standard borate solution are placed in a 50 ml volumetric flask and the solution made up to ca. 25 ml. Similarly 25 ml of the sample are pipetted into a 50 ml volumetric flask. 10 ml reagent solution are added to each of the solutions and their extinctions are measured at 414 nm in the photometer.

Interfering Factors

Iron ions at concentrations exceeding 5 mg/l can cause interference.

Calculation of Results

The boron concentration of the sample is determined from the calibration curve.

6.1.5 Calcium and Magnesium

Calcium and magnesium ions are present in all natural waters and are often cited as the cause of hardness. Hardness is defined as the ability of the water to cause precipitation of insoluble calcium and magnesium salts of higher fatty acids from soap solutions. Often only the calcium and magnesium concentrations in mol/l are given, rather than the hardness.

Calcium and magnesium carbonates play an important role in the formation of protective coatings in mains water pipes.

The complexometric method for determination of calcium and magnesium ions is described below, in which either calcium alone or both elements can be determined. The magnesium concentration can be calculated from the difference between the two measurements.

Areas of Application → water

a) Determination of Calcium

Apparatus

Titration equipment

Reagents and Solutions

EDTA solution:	3.725 g disodium EDTA ($C_{10}H_{14}N_2Na_2O_8 \cdot 2H_2O$) are made up to 1 l with water.
Indicator powder:	1 g calconcarboxylic acid ($C_{21}H_{14}N_2O_7S \cdot 3H_2O$) is ground with 99 g anhydrous sodium sulfate in a mortar.
Sodium hydroxide solution:	8 g sodium hydroxide are dissolved in 100 ml water.

Measurement

100 ml of the water sample are treated successively with 2 ml sodium hydroxide and 0.2 g indicator powder. Titration with EDTA solution is carried out rapidly until the colour changes from red to blue.

Interfering Factors

Barium and strontium ions are determined together with the calcium and magnesium ions. The colour change is unclear in the presence of some heavy metals.

Interference by iron and manganese ions in concentrations of up to ca. 5 mg/l can be almost entirely eliminated by the addition of 2 to 3 ml triethanolamine ($C_6H_{15}NO_3$).

Calculation of Results

The calcium content of the water sample is calculated as follows:

$$c\ (Ca^{2+}) = A \cdot C_E/B$$

c concentration of calcium ions in the sample, mmol/l
A volume of EDTA consumed, ml
B sample volume, ml
C_E concentration of the EDTA solution, mmol/l
   here: c (EDTA) = 0.01 mol/l = 10 mmol/l

b) Determination of both Calcium and Magnesium Ions

Apparatus

Titration equipment

Reagents and Solutions

EDTA solution:	as described under a)
Buffer solution, pH = 10:	6.75 g ammonium chloride and 0.05 g disodium magnesium EDTA ($C_{10}H_{12}N_2O_8Na_2Mg$) are dissolved in 57 ml ammonia, w (NH_3) = 25%, and made up to 100 ml with water.
Indicator solution:	0.5 g eriochrome black T are dissolved in 100 ml triethanolamine.

Sample Preparation

Suspended particles are removed by filtration through a 0.45 µm membrane filter before the analysis.

Measurement

100 ml of the sample are treated with 4 ml buffer solution and 3 drops of indicator solution. Titration with EDTA solution is carried out until the colour changes from red to blue. It is recommended that this first titration is carried out rapidly, because otherwise interference can occur due to carbonate precipitation.

The second titration is carried out by immediately adding ca. 0.5 ml less of the EDTA solution to 100 ml of the sample than were consumed in the first titration. After addition of buffer solution and indicator, titration is carried out until the colour changes from red to blue.

Interfering Factors
As described under a)

Calculation of results
The total content of magnesium and calcium ions in the water sample is calculated from:

$$c \, (Ca^{2+} + Mg^{2+}) = A/B \cdot C_E$$

c total concentration of calcium and magnesium ions, mmol/l
A volume of EDTA consumed, ml
B sample volume, ml
C_E concentration of the EDTA solution, mmol/l
     here: c (EDTA) = 0.01 mol/l = 10 mmol/l

c) Calculation of the Magnesium Content

To calculate the magnesium content of the sample, the result of measurement a) is subtracted from that of measurement b).

6.1.6   Calcium Carbonate Saturation and Equilibrium pH

The calcium carbonate precipitating or dissolving properties of water are principally determined by the chemical equilibrium between calcium carbonate, carbon dioxide and the ions constituting carbonic acid, besides the presence of other substances, such as magnesium, sulfate and chloride, their compounds and equilibrium constants. The equilibrium is important for the assessment of the corrosion of metallic and mineral materials and in water treatment. In accordance with the ordinance on drinking water of 1991 the pH of drinking water must be between 6.5 and 9.5. In addition it must not exceed the pH of calcium carbonate saturation. Several methods of determination with varying precision and requirements are in use: a) Tillman's curve, b) DIN 38 408-C 10-1 procedure, c) calculation of the calcium carbonate saturation from complex formation.

Areas of Application → water

a) Tillman's Curve

This process is still used for the approximate determination of the equilibrium pH (Fig. 31). The values for the acidic capacity to pH 4.3 ($K_{S\,4.3}$) and the base capacity to pH 8.2 ($K_{B\,8.2}$) are required. Using the approximation

$$K_{S\,4.3} \approx [HCO_3^-] \approx m$$
$$K_{B\,8.2} \approx [CO_2] \approx -p$$

the desired pH values are determined graphically. According to the definition, the curve is valid for a water temperature of 25 °C, an ionic strength of 0 mol/l and a value of $m-2[Ca^{2+}]$ of 0 mol/l.

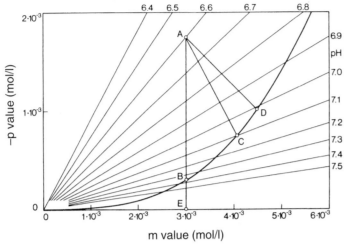

Fig. 31: Tillman's curve (Frimmel et al., 1993)

The abscissa represents the m value (corresponding approximately to the concentration of hydrogen carbonate ions) and the ordinate the negative p value (corresponding approximately to the CO_2 concentration). Waters whose m and $-p$ values are described by the curve drawn are in a state of calcium carbonate saturation at a particular pH. $m/-p$ co-ordinates lying above the equilibrium curve indicate calcium carbonate dissolving properties and those below the curve, calcium carbonate precipitating properties. There are three possible ways of de-acidifying a water sample, whose $m/-p$ values correspond to point A:

Degassing CO_2:

The m value does not change so that at point B vertically below A the calcium carbonate saturation is reached. The pH difference gives the saturation index SI. On further removal of CO_2 the water would become calcium carbonate precipitating, e.g. at point E.

Addition of alkali:

Sodium or calcium hydroxides are added to raise the pH. Since 1 mol hydrogen carbonate is formed for each mol carbon dioxide reacting, $m-p$ remains constant (note the plus or minus sign before p) and the straight line has a gradient of -1. The calcium carbonate saturation of the Tillman's curve is therefore reached at point C. Adding too much alkali can lead to precipitation of calcium carbonate.

Filtration through marble:

In this method 2 mol hydrogen carbonate are formed per mol carbon dioxide, i.e. $m-2p$ remains constant, the straight line has a gradient of $-1/2$ and meets the Tillman's curve at point D.

b) DIN 38 408-C 10-1 Procedure

For the approximate determination of the equilibrium pH, the water temperature, calcium concentration, acidic capacity and electrical conductivity must be known. At the pH known as the Langelier pH or pH_L the water would be in equilibrium with calcium carbonate without changing the hydrogen carbonate and calcium ion concentrations. The difference between pH_L and the pH of the water measured is known as the saturation index, SI.

The lowest possible saturation pH is given for 10 °C in Table 22 ($pH_{L,10}$). The correction factor a for a temperature t other than 10 °C is given by:

$$a = 0.01 \cdot (10{-}t).$$

pH_L is then calculated according to:

$$pH_L = pH_{L,10} + a$$

The table cannot be used at high chloride or sulfate concentrations, sometimes indicated by increased electrical conductivity. The values for $pH_{L,10}$ are given in the upper line of the table and below them the values for the electrical conductivity K_{M10} for a standard water sample at 10 °C in μS/cm. The table can be used when $K_{25} < 2.1 \cdot K_{M10}$ and $K_{10} < 1.5 \cdot K_{M10}$. To calculate the conductivity at 10 °C the factors listed on the right hand side are used.

Tabelle 22: pH values of the calcium carbonate saturation ($pH_{L,\,10}$) of water at 10 °C as a function of acidic capacity and calcium concentration

Ca in mg/l	Ca in mmol/l	\multicolumn{15}{c}{Acidic capacity up to pH 4.3 (≡ m value) in mmol/l}														t (°C)	f_{25}
		0.25	0.5	0.75	1	1.25	1.5	2	2.5	3	3.5	4	5	6	7	8		
10	0.25	9.65	9.18	8.98	8.84												0	1.918
		54	59	75	91												1	1.857
																	2	1.800
20	0.5	9.28	8.88	8.69	8.54	8.44	8.36	8.24									3	1.745
		112	106	104	119	135	150	181									4	1.693
																	5	1.643
40	1.0	9.00	8.61	8.41	8.27	8.17	8.08	7.96	7.87	7.80							6	1.596
		222	216	210	204	198	204	234	264	294							7	1.551
																	8	1.508
50	1.25	8.92	8.53	8.32	8.19	8.08	8.00	7.88	7.78	7.71	7.65	7.60					9	1.467
		276	270	264	258	252	246	260	290	319	349	379					10	1.428
																	11	1.390
60	1.5	8.85	8.46	8.26	8.12	8.02	7.93	7.81	7.71	7.64	7.58	7.52	7.44				12	1.354
		330	323	317	311	305	299	286	315	345	374	403	461				13	1.320
																	14	1.287
80	2.0	8.74	8.36	8.16	8.02	7.92	7.83	7.70	7.61	7.53	7.47	7.41	7.33	7.26	7.20		15	1.256
		435	428	422	416	409	403	390	378	394	423	451	509	566	624		16	1.225
																	17	1.196
100	2.5	8.67	8.28	8.08	7.94	7.84	7.76	7.63	7.53	7.45	7.38	7.33	7.24	7.17	7.11	7.06	18	1.168
		538	531	525	518	512	505	492	480	467	470	499	556	612	669	725	19	1.141
																	20	1.116
120	3			8.02	7.88	7.78	7.70	7.57	7.47	7.39	7.32	7.26	7.17	7.10	7.04	7.00	21	1.091
				626	619	613	606	593	580	567	554	545	601	657	713	769	22	1.067
																	23	1.044
160	4				7.79	7.69	7.60	7.47	7.37	7.29	7.23	7.17	7.07	7.00	6.94	6.89	24	1.021
					818	811	805	791	778	765	751	738	711	745	801	856	25	1.000
																	26	0.979
200	5						7.53	7.40	7.30	7.22	7.16	7.10	7.00	6.92	6.86	6.81	27	0.959
							999	986	972	958	945	931	904	877	885	940	28	0.940
																	29	0.921
280	7							7.20	7.12	7.05	7.00	6.90	6.82	6.75	6.70		30	0.903
								1351	1337	1323	1308	1280	1252	1224	1196			

c) Calculation of the Calcium Carbonate Saturation from Complex Formation

There are several commercially available calculation programs for the exact calculation taking into account the compounds of the various constituents present in water as well as the carbonate and sulfate complexes of calcium and magnesium. Most are based on the chemical relationships described in DIN 38 408-C 10-1. These are interactive programs with menus which give, besides the actual calculation, a range of options for input and output of data. Available water data can be stored as data files and be called up from the program when required.

The possibilities for calculating data on calcium carbonate saturation and parameters for water treatment are shown in Table 23. The program (here BWASATW2) thus checks the plausibility of various measured values, determines auxiliary parameters, such as ionic strength and the *m* value, and then calculates the calcium carbonate saturation data from the most favourable of the three variants.

Table 23: Values measured for a drinking water sample and data calculated for calcium carbonate saturation

Parameter measured	Data measured	Observations
Temperature, °C	9.6	
Electrical conductivity, µS/cm	384	(at 25 °C)
pH value	7.30	
Acidic capacity, mmol/l	3.27	(at 25 °C)
Base capacity, mmol/l	0.38	
Sodium, mg/l	6.90	
Potassium, mg/l	1.96	
Calcium, mg/l	60.12	
Magnesium, mg/l	12.15	
Chloride, mg/l	12.4	
Sulfate, mg/l	28.8	
Nitrate, mg/l	9.30	
Hardness, mmol/l	2.00	
Data on the water data calculated from the above		
Ionic strength, mmol/l	6.38	
m value	3.25	(calculated from pH and Σ strong electrolytes)
Total carbonic acid, mmol/l	3.60	
Charge balance, mmol/l	0.00	
Data calculated for calcium carbonate saturation		
pH value	7.39	
pH_L	7.61	(Langelier pH)
BI, mmol/l	0.75	(buffering intensity = $\Delta m / \Delta pH$)

Table 23: (Continued)

Parameter measured	Data measured	Observations
SI_c	− 0.22	(calcium carbonate saturation index at the calculated temperature)
pH_c	7.55	(pH after setting up the equilibrium with calcium carbonate)
D_c, mmol/l	0.103	(addition or removal of calcium carbonate until pH_c is reached)
pH_s (CaO)	7.58	(saturation pH after adjustment with CaO)
D (CaO) mmol/l	0.061	(addition of CaO for $SI_c = 0$)
pH_s (NaOH)	7.59	(saturation pH after adjustment with NaOH)
D (NaOH), mmol/l	0.130	(addition of NaOH for $SI_c = 0$)
pH_a (CO_2)	7.61	(saturation pH after blowing out CO_2)
D (CO_2), %	− 41.6	(quantity of CO_2 blown out for $SI_c = 0$)
CO_2 bound, mmol/l	0.212	('bound carbonic acid' according to Tillman)
CO_2 corrosive, mmol/l	0.141	('corrosive carbonic acid' according to Tillman)
Ca^{2+}/Ca_{tot}, %	95.13	(proportion of free calcium)
SI (atm CO_2)	1.27	('saturation index' of CO_2, i.e. logarithm of the quotient of the calculated equilibrium CO_2 partial pressure and the atmospheric partial pressure (0.35 hPa))

6.1.7 Chemical Oxygen Demand

The chemical oxygen demand (COD) is defined as the quantity of oxygen consumed in the oxidation of a sample under defined conditions. The extent of oxidation depends on the nature of the oxidisable substances, the pH value, the temperature, the reaction time, the concentration of the oxidising agent and the type of added accelerators.

Potassium permanganate has been used for a long time as an oxidising agent for the determination of organic compounds in water and waste water. The procedure is relatively easy to carry out but suffers from the disadvantage that substances, such as some amino acids, ketones or saturated carboxylic acids, are not, or only partially, oxidised. Therefore the method is only used for slightly contaminated surface water or for preliminary testing of drinking water. The reaction takes place in acidic solution:

$$MnO_4^- + 8H^+ + 5e^- \rightarrow Mn^{2+} + 4H_2O$$

The COD determination with potassium dichromate is particularly important for testing waste water, because according to the Waste Water Charges Act in Germany the COD also determines the discharge level. This process gives a more exact determination of the COD than the oxidation with potassium permanganate. The reaction is as follows:

$$Cr_2O_7^{2-} + 14H^+ + 6e^- \rightarrow 2Cr^{3+} + 7H_2O$$

With this oxidising agent all organic substances are almost completely oxidised, with few exceptions. Concentrations of ca. 10 to 15 mg/l, expressed as oxygen $\beta(O_2)$, can be determined with certainty. Certain inorganic ions (e.g. nitrite, sulfite and Fe(II) ions) are also oxidised.

For the assessment of the results it is important to note that the COD does not permit the calculation of quantities of organic substances, whose proportions in the sample are unknown. Different substances require different quantities of oxidising agent for complete oxidation. Some examples are:

oxalic acid ($C_2H_2O_4$): 0.18 mg O_2 per mg substance
acetic acid ($C_2H_2O_2$): 1.07 mg O_2 per mg substance
phenol (C_6H_6O): 2.38 mg O_2 per mg substance

For domestic waste water m (COD) = 1.2 mg per mg organic material is often used.

Besides the description of the potassium permanganate consumption and the DIN-COD (dichromate procedure), a COD cuvette test will be described, which is frequently used in plant laboratories.

Areas of Application ➜ water, waste water, soil (see Section 6.3.1.4)

a) Potassium Permanganate Consumption

Apparatus
Titration equipment

Reagents and Solutions

Potassium permanganate solution:	3.1608 g potassium permanganate are made up to 1 l with freshly distilled water. 100 ml of this solution are taken and made up to 1 l. The titration factor of the solution is always determined with oxalic acid before use. The solutions are stored in dark bottles.
Oxalic acid solution:	6.3033 g oxalic acid ($C_2H_2O_4 \cdot 2H_2O$) are made up to 1 l with freshly distilled water and 50 ml conc. sulfuric acid. 100 ml of this solution are made up to 1 l with water and 50 ml conc. sulfuric acid. The solution is stable for up to 6 months in dark bottles.
Sulfuric acid, w (H_2SO_4) = 36%:	100 ml conc. sulfuric acid are added carefully to 200 ml water. KMnO$_4$ solution is added to this solution while still hot until a pink colour appears.

Sample Preparation

The measurements should be carried out as soon as possible after sample collection. The glass apparatus employed for boiling should be stored under dust-free conditions.

Measurement

The determination of the titration factor of the $KMnO_4$ solution is carried out by heating to boiling 100 ml distilled water and 5 ml sulfuric acid, w (H_2SO_4) = 36%, together with antibumping granules. $KMnO_4$ solution is added until a slight pink coloration appears. Then 20 ml oxalic acid are added and the solution again titrated with $KMnO_4$ solution until slightly pink. The consumption x should lie between 19 and 21 ml. The titration factor t is calculated according to:

$$t = 20/x.$$

The measurement is performed by placing 100 ml of the sample (or a smaller volume made up to 100 ml with water) in an Erlenmeyer flask followed by 5 ml sulfuric acid w (H_2SO_4) = 36%. The flask is covered with a cold finger or watch glass and brought to boiling in ca. 5 min. While boiling, 20 ml $KMnO_4$ solution is added from a pipette and the solution allowed to simmer for 10 min. Then 20 ml oxalic acid solution are added and the mixture heated until the colour disappears completely. The hot solution (ca. 80 °C) is then titrated with $KMnO_4$ solution until a pink colour lasts for ca. 30 sec. The volumes consumed should lie between 4 and 12 ml. If they are higher or if the sample was already colourless before addition of oxalic acid, the determination is repeated with a smaller sample volume. A blank with 100 ml dilution water is measured in parallel.

Interfering Factors

Hydrogen sulfide, sulfides and nitrites interfere, but are removed in an acidic medium. Chloride ion concentrations exceeding 300 mg/l can interfere. In such cases the sample is diluted.

Calculation of Results

The $KMnO_4$ consumption β ($KMnO_4$) is given in mg/l as follows:

$$\beta (KMnO_4) = (a-b) \cdot t \cdot 316 \text{ mg}/V$$

a $KMnO_4$ consumed by sample, ml
b $KMnO_4$ consumed by blank, ml
t titration factor of the $KMnO_4$ solution
V sample volume, ml

b) Potassium Dichromate Consumption (Rapid Test according to DIN 38 409, part 43)

Apparatus

Cylindrical reaction vessel (e.g. 30 cm high, 3 cm in diameter) with ground glass joint and air condenser
Aluminium heating plate or heating bath, thermostatable
Titration equipment

Reagents and Solutions

Silver sulfate solution:	80 g silver sulfate are dissolved in 1 l conc. sulfuric acid.
Potassium dichromate solution:	4.9031 g potassium dichromate ($K_2Cr_2O_7$) (dried for 2 h at 105 °C) are made up to 1 l with water.

Ferrous ammonium sulfate solution: 98 g ferrous ammonium sulfate ($Fe(NH_4)_2(SO_4)_2 \cdot 6H_2O$) are dissolved in water. 20 ml conc. sulfuric acid are added and the solution made up to 1 l with water. The titration factor of this solution is determined with $K_2Cr_2O_7$ solution before use.

Ferroin indicator solution: 0.98 g ferrous ammonium sulfate and 1.485 1,10-phenanthroline ($C_{12}H_8N_2 \cdot H_2O$) are made up to 100 ml with water.

Mercury(II) sulfate solution: 15 g mercury(II) sulfate are dissolved in water. After addition of 10 ml conc. sulfuric acid, the solution is made up to 100 ml with water. The solution is suitable for masking chloride concentrations of up to 1000 mg/l.

Sample Preparation

The measurements should be performed as soon as possible after sample collection. Glass equipment should be kept under dust-free conditions and ground glass joints must not be greased. Waste water samples are usually tested after being left for 2 h to remove solids by sedimentation.

Calibration and Measurement

To determine the titration factor of the ferrous ammonium sulfate solution 10 ml of the conc. potassium dichromate solution are diluted to 100 ml. 30 ml conc. sulfuric acid are added, the solution cooled and titrated against ferrous ammonium sulfate solution after addition of 3 drops of ferroin indicator. The titration factor t is calculated from the consumption x (in ml) according to:

$$t = 10/x.$$

The titration factor should be checked daily. The determination is carried out as follows: 20 ml of the sample (or an aliquot diluted to 20 ml), 10 ml potassium dichromate solution and 2.5 ml mercury sulfate solution are mixed. Some antibumping granules and 40 ml conc. sulfuric acid are added via the condenser. The mixture is heated to boiling. 5 min after boiling has begun 5 ml silver sulfate solution are added via the condenser and boiling is continued for 10 min. The mixture is cooled for 5 min at air temperature and 50 ml water are added carefully via the condenser. The mixture is then left to cool to room temperature in the water bath. After addition of 3 drops of ferroin indicator the excess potassium dichromate is back-titrated with ferrous ammonium sulfate solution. At the end point the indicator changes from blue-green to red-brown. A blank sample, consisting of 20 ml dilution water, is treated in the same way as the sample.

If the chloride concentration in the sample is less than 100 mg/l, addition of mercury sulfate solution is unnecessary.

The quality of the method can be checked using potassium phthalate. 425.1 mg potassium phthalate (dried at 105 °C) are made up to 1 l with water. This solution has a COD set value of $\beta(O_2)$ = 500 mg/l.

Interfering Factors

Interference caused by chloride ions can be prevented as described above.

Calculation of Results

The COD, expressed as oxygen β (O_2) (in mg/l), is calculated according to the following equation:

$$\beta(O_2) = \frac{(a-b) \cdot t \cdot 2000 \, mg}{V}$$

a   consumption of ferrous ammonium sulfate solution for the blank value, ml
b   consumption of ferrous ammonium sulfate for the sample, ml
t   titration factor of the ferrous ammonium sulfate solution
V   sample volume, ml

c)   COD Cuvette Test (e.g. Drs Lange, Macherey and Nagel)

The measurement of the COD using a cuvette test uses the different absorption maxima of Cr(VI) (yellow) and Cr(III) (green).

The digestion is carried out under the same conditions as in DIN 38409, part 41, i.e. 2 h at 148 °C. The solution is measured at 620 nm in the photometer. The advantage of the cuvette test principally lies in the low consumption of the hazardous substances mercury, silver and sulfuric acid. Used cuvettes are taken back by the producer and recycled.

6.1.8   Chloride

Chloride is present in all natural waters at greatly varying concentrations, depending on the geochemical conditions. Particularly high concentrations occur in water near salt deposits. Large amounts of chloride reach waste water through faecal discharge. For this reason chloride can serve as a pollution indicator when considered together with other parameters and when a natural geological origin does not apply. A concentration of even 250 mg/l can give a salty taste to drinking water with low mineral content, whereas in the presence of larger quantities of calcium and magnesium ions a concentration of ca. 1000 mg/l is necessary for this.

The volumetric determination with silver nitrate using potassium chromate as the end point indicator is described below.

Areas of Application → water, waste water, soil

Apparatus

Titration equipment

Reagents and Solutions

Silver nitrate solution:	4.791 g silver nitrate are made up to 1 l with water. The solution is stored in a brown glass bottle. 1 ml corresponds to 1 mg chloride ions.
Sodium chloride solution:	1.648 g sodium chloride (dried for 2 h at 105 °C) are made up to 1 l with water. 1 ml of this solution contains 1 mg chloride ions.
Potassium chromate solution:	10 g potassium chromate are dissolved in water and the solution made up to 100 ml.

Sample Preparation

At pH values below 5 a small amount of calcium carbonate is added and the sample shaken. At pH values exceeding 9.5 the sample is first titrated against sulfuric acid, c (H_2SO_4) = 0.1 mol/l, with phenolphthalein as indicator. The quantity of acid required is then added to a second sample together with some calcium carbonate.

Measurement

100 ml of the sample (or a smaller sample at higher chloride concentrations) are placed in an Erlenmeyer flask with 1 ml potassium chromate. The mixture is titrated against a white background with silver nitrate solution until the colour changes from greenish-yellow to reddish-brown.

A blank sample with distilled water is treated in the same way as the sample.

Interfering Factors

Bromide, iodide or cyanide are co-determined by this method. Sulfite, sulfide and thiosulfate can be removed as follows:

After careful acidification of the sample with sulfuric acid, it is boiled for a few min. 3 ml hydrogen peroxide, w (H_2O_2) = 10%, are added and boiling continued for 15 min. The losses through evaporation are then compensated for. Sodium hydroxide, c (NaOH) = 1 mol/l, is added dropwise until the mixture is weakly alkaline. It is boiled again for a short time, filtered and the analysis then performed as described above.

Discoloured water samples can be decolourised with activated charcoal and then filtered.

Calculation of Results

The chloride concentration β (Cl⁻), given in mg/l, is calculated according to the following equation:

$$\beta\,(\text{Cl}^-) = (A\!-\!B) \cdot 1000 \text{ mg} \cdot \text{ml}^{-1}/C$$

A   volume of silver nitrate solution consumed for the sample, ml
B   volume of silver nitrate solution consumed for the blank, ml
C   sample volume, ml

6.1.9   Chromate

As salts of chromic acid (H_2CrO_4), chromates are important industrial chemicals. On acidification of chromate-containing solutions dichromates are first formed and under strongly acidic conditions, polychromates. Chromium is hexavalent in these compounds. Dissolved Cr(VI) is reduced to sparingly soluble Cr(III) by many organic compounds.

The mammalian toxicity of Cr(VI) compounds is 100 to 1000 times higher than that of Cr(III) compounds. Some soluble chromates are mutagenic.

The analysis is based on DIN 38 405, part 24. It can be used to determine Cr(VI) concentrations of up to 3 mg/l. The procedure involves the oxidation of 1,5-diphenylcarbazide to 1,5-diphenyl-carbazone, which forms a red dye complex with chromium.

Areas of Application ➔ water, waste water

Apparatus
Spectrophotometer or fixed filter photometer with 550 nm filter
pH meter with electrode
Membrane filtration apparatus with 0.2 μm filters

Reagents and Solutions

Buffer solution:	456 g dipotassium hydrogen phosphate ($K_2HPO_4 \cdot 3H_2O$) are dissolved in 1 l water and the pH adjusted to 9.0.
Aluminium sulfate solution:	247 g aluminium sulfate ($Al_2(SO_4)_3 \cdot 18H_2O$) are dissolved in 1 l water.

Diphenylcarbazide solution:	1 g 1,5-diphenylcarbazide is dissolved in 100 ml acetone. After addition of 1 drop of glacial acetic acid, the solution is transferred to a brown bottle and can be kept for ca. 2 weeks at 4 °C.
Chromium(VI) solutions:	2.829 g potassium dichromate are dissolved in water and made up to 1 l in a volumetric flask. 10 ml of this stock solution are made up to 1 l with water in a second volumetric flask.

Phosphoric acid, w (H_3PO_4) = 87%

Calibration and Measurement
10 ml buffer is added to 1 l of the sample immediately after collection. The pH is measured and adjusted to 7.5 to 8.0 with sodium hydroxide or phosphoric acid as appropriate. Then 1 ml aluminium sulfate is added, the mixture shaken and the pH adjusted to 7.0 to 7.2 with phosphoric acid.

100 ml of the pretreated sample are filtered through a membrane filter. After discarding the first few ml, 50 ml are transferred to a 100 ml volumetric flask (less at higher Cr(VI) concentrations). 2 ml phosphoric acid and 2 ml diphenylcarbazide solution are added and the solution made up to the mark with water. After letting the solution stand for 5 to 10 min, the extinction is measured at 550 nm against water.

A blank sample with water is treated in the same way as the sample. A calibration curve is drawn using measurements taken of the extinctions of chromate solutions with concentrations within the required measurement range, which have been treated in the same way but not filtered.

Interfering Factors

It is important to test the samples quickly after collection. Oxidising or reducing agents can cause interference. In the presence of oxidising agents 1 ml 10% sodium sulfite solution is added at the end of the pretreatment stage and sulfite paper is used to test whether an excess is present. In the second step described above 1 ml sodium hypochlorite solution (ca. 150 g/l Cl_2) is added immediately to remove the excess of sulfite and other reducing agents. 10 g sodium chloride are then added and air is passed through the sample for 40 min (use local extraction!).

If the sample has an inherent coloration after filtration, a second filtered and pretreated sample is processed in the same way but without the addition of diphenylcarbazide and the extinction measured at 550 nm is subtracted from the first measurement.

Nitrite concentrations of > 20 mg/l and very high ammonium concentrations also cause interference.

6.1.10   Cyanides

Cyanides are used in some industrial and manufacturing processes (hardening shops, electroplating plants) or are liberated (coking plants). They can reach waste water or bodies of water as a result of incidents.

Cyanide ions are toxic towards water organisms, whereas many complexed cyanides are only slightly toxic. Cyanide ions can be liberated from complex cyanides through rearrangements so that assessing the toxic properties of water samples without the use of biological tests is often difficult.

Conventionally a distinction is made between 'total cyanide' and 'easily released cyanide', according to DIN 38 405, part 13. Other terms, such as 'free cyanide' or 'cyanide which can be decomposed with chlorine', should be avoided. Total cyanide is the sum of simple and complex cyanides which can release hydrogen cyanide under the conditions of the process. 'Easily released cyanides' are those compounds with CN groups which release hydrogen cyanide at room temperature and pH 4.

The presence of cyanides in ground water, surface water and drinking water indicates a discharge of waste water or seepage from a dump and should always be followed up with an investigation and remedial measures. In crude water for drinking water processing and in drinking water itself, cyanide concentrations may not exceed 0.05 mg/l.

Areas of Application ➔ water, waste water

a) Total Cyanide

Apparatus
Distillation equipment with 500 ml three-necked flask, reflux condenser, absorption vessel (Fig. 32), funnel, water pump, wash bottle, flow meter, heater
Spectrophotometer or fixed filter photometer with 580 nm filter

a   **b**

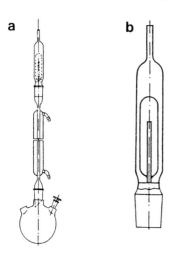

Fig. 32: (a) Distillation apparatus, (b) absorption vessel

Reagents and Solutions

Tin(II) chloride solution:	50 g tin chloride ($SnCl_2 \cdot 2H_2O$) are dissolved in 40 ml hydrochloric acid, c (HCl) = 1 mol/l, and the solution made up to 100 ml with water.
Copper sulfate solution:	20 g copper sulfate ($CuSO_4 \cdot 5H_2O$) are dissolved in water and the solution made up to 100 ml.
Chloroform – phenolphthalein solution:	0.03 g phenolphthalein are dissolved in 90 ml ethanol and then treated with 10 ml chloroform.
Zinc-cadmium sulfate solution:	10 g zinc sulfate ($ZnSO_4 \cdot 7H_2O$) and 10 g cadmium sulfate ($3CdSO_4 \cdot 8H_2O$) are made up to 100 ml with water.
Buffer solution, pH = 5.4:	6 g sodium hydroxide are dissolved in 50 ml water and treated with 11.8 g succinic acid ($C_4H_6O_4$). The solution is made up to 100 ml with water.
Chloramine-T solution:	1 g chloramine-T ($C_7H_7SO_2NClNa$) is dissolved in 100 ml water. The solution is stable for ca. 1 week.
Barbituric acid – pyridine reagent:	3 g barbituric acid ($C_4H_4N_2O_3$) are placed in a 50 ml volumetric flask, mixed with a little water and treated with 15 ml

	pyridine. Water is added with shaking until the barbituric acid completely dissolves. Then 3 ml hydrochloric acid, w (HCl) = 25%, is added followed by water up to the mark. The reagent can be kept for 1 week in the refrigerator.
Standard cyanide solution	β (CN$^-$) = 10 mg/l: 25 mg potassium cyanide are dissolved in 1 l sodium hydroxide, c (NaOH) = 1 mol/l.
Hydrochloric acid:	w (HCl) = 25% and c (HCl) = 1 mol/l
Sodium hydroxide:	w (NaOH) = 20% and c (NaOH) = 1 mol/l

Sample Preparation

After collection 5 ml sodium hydroxide, c (NaOH) = 1 mol/l, 10 ml chloroform – phenolphthalein solution and 5 ml tin(II) chloride solution are added to 1 l of the sample. If there is a red colouration, hydrochloric acid, c (HCl) = 1 mol/l, is added until the colour disappears. If there is no red coloration sodium hydroxide, c (NaOH) = 1 mol/l, is added dropwise until a red coloration appears. 10 ml zinc-cadmium sulfate solution are added and the sample is stored in a cool, dark location until testing. Before testing a portion of the sample the container should be shaken well.

Calibration and Measurement

10 ml sodium hydroxide, c (NaOH) = 1 mol/l, are placed in the absorption vessel and the latter connected to the water pump. The air sucked through is first passed through a wash bottle filled with 100 ml sodium hydroxide, c (NaOH)= 1 mol/l.

The following are placed successively in the three-necked flask: 30 ml demineralised water, 10 ml copper sulfate solution, 2 ml tin chloride solution, 100 ml water sample (or an aliquot made up to 100 ml) and 10 ml hydrochloric acid, w (HCl) = 25%. Air is sucked at ca. 20 l/h through the apparatus and then the contents of the flask are boiled gently for 1 h. To determine the cyanide ion concentration the contents of the absorption vessel are then transferred to a 25 ml volumetric flask, the vessel rinsed into the flask and the latter filled with water up to the mark. 10 ml are removed from this flask and transferred to another 25 ml volumetric flask. 2 ml buffer solution, 4 ml hydrochloric acid, c (HCl) = 1 mol/l, and 1 ml chloramine-T solution are also placed in this flask and the contents allowed to stand for 5 min. After addition of 3 ml barbituric acid – pyridine reagent, water is added up to the mark. After 20 min the extinction of the solution is measured at 580 nm against a control solution, which contains the substances named in the same concentrations. A blank sample with distilled water is subjected to the whole determination procedure in the same way as the sample.

Calculation of Results

The concentration of the total cyanide in the water sample, given in mg/l, is calculated according to the following equation:

$$\beta \text{(total cyanide)} = (A–B) \cdot 2500/e \cdot V$$

A   cyanide content of the sample, mg
B   cyanide content of the blank, mg
e   factor accounting for volume increase through addition of preservative, $e = 0.97$
V   sample volume, ml

b) Easily Released Cyanide

Apparatus
Distillation equipment, as described under a)

Reagents and Solutions

Buffer solution, pH = 4:   80 g potassium hydrogen phthalate ($C_8H_5O_4K$) are dissolved in 920 ml warm water.

EDTA solution: 100 g disodium EDTA ($C_{10}H_{14}N_2O_8Na_2 \cdot 2H_2O$) are dissolved in 940 ml warm water.

The other solutions required are prepared as in a).

Sample Preparation
As described under a)

Calibration and Measurement
10 ml sodium hydroxide, c (NaOH)= 1 mol/l, are placed in the absorption vessel and a current of air is passed through at 30 to 60 l/h. 10 ml zinc-cadmium sulfate solution, 10 ml EDTA solution, 50 ml buffer solution and 100 ml of the shaken sample are added. The pH is adjusted to 4 with hydrochloric acid. The flask is connected to a wash bottle filled with 100 ml sodium hydroxide, c (NaOH)= 1 mol/l, and air is sucked through the solution at 60 l/h for 4 h. A blank is treated in the same way as the sample.

The calibration, measurement and calculation of results are as described under a).

Interfering Factors
Interference can occur in the presence of high concentrations of sulfide, sulfite, thiosulfate, carbonate, nitrate, nitrite and ammonium.

Copper ions accelerate the decomposition of any hexacyanoferrate which may be present and inhibit the transfer of hydrogen sulfide to the absorption vessel. The addition of tin(II) reduces Cu(II) to Cu(I) and also inhibits the formation of undesired cyanogen chloride or cyanates. The decomposition of hexacyanoferrates (II and III) is also complete at high iron(II) and iron(III) concentrations. The zinc and cadmium sulfates stabilise hexacyanoferrates against partial decomposition and act as sulfide acceptors. The addition of EDTA leads to the liberation of hydrogen cyanide from cyanonickel complexes. The order of addition of the solutions given above should be adhered to.

6.1.11   Density

The density of water reaches a maximum at 4 °C and an air pressure of 1013 hPa. On warming above 4 °C and between 4 and 0 °C the water expands. The nature and quantity of substances dissolved or suspended in water can alter the density. In drinking water the deviations are normally ignored, but not in mineral waters, ground water or waste water, which can contain large quantities of dissolved substances. If the volume rather than the weight of a water sample is measured before examination, it should be multiplied by the density.

Areas of Application → water, waste water

a) Pyknometric Determination

Apparatus
Pyknometer of 20 to 100 ml capacity with narrow, graduated neck
Thermostat

Sample Preparation
Undissolved substances are removed by filtration through a fluted filter paper.

Measurement
The sample is placed in the dry pyknometer free of air bubbles approximately up to the mark and is tempered at 20 °C in a thermostat for ca. 1 h. Water is then added up to the mark. The vessel, which must be dry on the outside, is weighed accurately to 1 mg, emptied, cleaned and filled with water at 20 °C up to the mark. It is kept at 20 °C for a short time and reweighed.

Calculation of Results
The density of the liquid D_{20} (g/cm^3) is calculated as follows:

$$D_{20} = \frac{(G_1 - G_E) \cdot 0.99820 \text{g} \cdot \text{cm}^{-3}}{G_2 - G_E}$$

G_1 weight of the vessel filled with the sample, g
G_2 weight of the vessel filled with water, g
G_E weight of the empty vessel, g
0.99820 = density of water at 20 °C, g · cm^{-3}

For densities of < 1 the result is given to 4 decimal places and for those > 1, to 3 decimal places. At temperatures other than 20 °C the corresponding correction factor is used (Table 24).

Table 24: Density of water as a function of temperature

Temperature (°C)	Density ρ (kg/l)	Temperature (°C)	Density ρ (kg/l)
0	0.99984	20	0.99820
5	0.99996	25	0.99704
10	0.99970	30	0.99564
15	0.99910	35	0.99403

b) Determination with the Mohr-Westphal Balance

In the density determination using the Mohr-Westphal balance a weighted body of known volume V (usually 1 cm^3) is attached to a holding device with a fine wire and its weight is balanced by counterweights on the other side of the balance arm. After immersing the body in water kept at 20 °C the apparent weight loss caused by the upthrust is compensated for by removing a weight from the balance arm. Figure 33 shows the important components of the Mohr-Westphal balance.

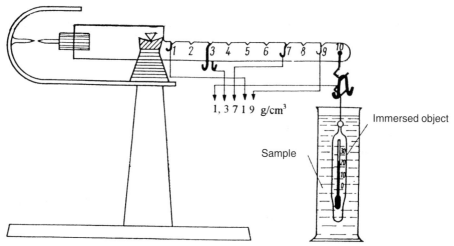

1, 3 7 1 9 g/cm^3

Sample

Immersed object

Fig. 33: Mohr-Westphal balance

Measurement
The sample kept at 20 °C is transferred to the measuring cylinder. The weighted body is then hung on the balance hook, thus displacing ca. 1.5 cm liquid. Then the 1 g rider is hung directly on to the hook above the weighted body. If the weight on the weighted body side is too small, more riders are placed on the balance between the first and ninth notches until equilibrium is reached. The individual weights are then added:

- If there is equilibrium on hanging a 1 g rider on the hook, for a weighted body volume of 1 cm^3 the density D_{20} is 1 g/cm^3.
- Hanging a 1 g rider between the first and ninth notch gives the first decimal place, i.e. the notch number $\cong$ the number after the point.
- The same applies for 0.1 g and 0.01 g riders.

Calculation of Results
The density at 20 °C is calculated from

$$D_{20} = G \cdot 0.99820/V$$

c) Areometric Determination

The areometric density determination is carried out at 20 °C by immersing a precision areometer into the water sample contained in a measuring cylinder. The areometer should be dry and free of grease. The density is read directly on the areometer.

6.1.12 Iron (Total Iron and Iron(II))

Under anaerobic conditions the content of Fe^{2+} can be several mg/l, particularly in ground water and waste water, while in most aerobic surface waters the concentration seldom exceeds 0.3 mg/l. Fe^{2+} ions are oxidised fairly rapidly by atmospheric oxygen. First yellow-brown colloidal iron(III) hydroxide is formed, which then precipitates as the brown hydroxide.

$$2Fe^{2+} + 1/2O_2 + 2H^+ \rightarrow 2Fe^{3+} + H_2O$$

Iron is an undesirable component of drinking and industrial water because iron hydroxide can form deposits in pipes and cause problems in usage at higher concentrations (metallic taste, stains on textiles after washing).

In the determination of iron a distinction is made between total iron and iron(II). In addition, a distinction is made between total iron, i.e. the total of dissolved and undissolved iron, and the total dissolved iron, i.e. the sum of the iron(II) and iron(III) compounds.

Areas of Application ➔ water, waste water

a) Total Iron

Apparatus
Spectrophotometer or fixed filter photometer with 520 nm filter

Reagents and Solutions

2,2'-Dipyridyl solution:	0.1 g 2,2'-dipyridyl ($C_{10}H_8N_2$) are made up to 100 ml with water.
Standard iron(III) solution, β_1 (Fe^{3+}) = 100 mg/l, β_2 (Fe^{3+}) = 10 mg/l:	100 mg iron wire are dissolved in 20 ml H_2O and 5 ml hydrochloric acid, w (HCl) = 36%, with warming in a 1 l volumetric flask. After cooling the volume is made up to 1 l with water. 100 ml of this solution is made up to 1 l with water. The diluted solution should be freshly prepared daily.

Hydrochloric acid, w (HCl) = 25%
Nitric acid, w (HNO$_3$) = 65%
Sodium acetate ($C_2H_3O_2Na \cdot 3H_2O$)
Ascorbic acid ($C_6H_8O_6$)

Sample Preparation

For the determination of total iron, the pH of the sample is adjusted to pH 1 with hydrochloric acid at the sampling point.

If the total dissolved iron is to be determined, the sample is filtered immediately after collection through a medium-coarse filter paper and the filtrate acidified to pH 1.

Sparingly soluble iron compounds must be removed: 5 ml nitric acid, w (HNO_3) = 65%, and 10 ml hydrochloric acid, w (HCl) = 25%, are added to 50 ml of the sample in a beaker and the contents heated to ca. 80 °C until the solid particles dissolve. 2 ml conc. sulfuric acid are then added and the solution concentrated until sulfuric acid fumes appear. After cooling 20 ml of water are added, the mixture is filtered if necessary and made up to 100 ml with water in a volumetric flask.

Calibration and Measurement

Different quantities of the standard solution are placed in a series of volumetric flasks to provide a range of Fe(III) concentrations from 0.05 to 1 mg/l. These solutions are treated in the same way as the sample itself.

To determine the total iron in the samples 50 ml (or a smaller volume) are pipetted into a 100 ml volumetric flask and the pH adjusted to 2 to 3 if necessary. After addition of 300 to 500 mg ascorbic acid to reduce Fe(III) to Fe(II), 20 ml 2,2'-dipyridyl solution are added. The solution is buffered to pH 5–6 by adding 2 to 5 g solid sodium acetate and the volume made up to 1 l. The extinction is measured at 520 nm in the photometer. A blank is treated in the same way as the sample.

Interfering Factors

Copper, zinc and orthophosphate can cause interference if their concentrations are ten times greater than that of iron.

Cyanides are usually removed by the addition of acid. If the presence of complex cyanides is suspected, the digestion procedure described for their removal should be followed.

Calculation of Results

The content of total iron is determined from the calibration curve, taking the blank value into consideration.

b) Iron(II)

Apparatus

Spectrophotometer or fixed filter photometer with 520 nm filter

Reagents and Solutions

2,2'-Dipyridyl solution:	as described under a)
Standard iron(II) solution,	0.7022 g ferrous ammonium sulfate
β_1 (Fe^{2+}) = 100 mg/l,	((NH_4)$_2$Fe(SO_4)$_2$ · $6H_2O$) are made up to 1 l with water.
β_2 (Fe^{2+}) = 10 mg/l:	100 ml of this solution are made up to 1 l with water. Both
	solutions must be freshly prepared daily.

Sodium acetate ($C_2H_3O_2Na$ · $3H_2O$)

Sample Preparation

Before sample collection 20 ml 2,2'-dipyridyl solution are placed in a 100 ml volumetric flask. Depending on the expected Fe(II) content, 10 to 75 ml of sample are pipetted into the volumetric flask at the sampling point.

Calibration and Measurement

Several 20 ml portions of 2,2'-dipyridyl solution are placed in a series of 100 ml volumetric flasks. Different quantities of fresh standard solution are placed in these flasks in order to set up a range of Fe(II) concentrations from 0.05 to 1 mg/l. The solutions are made up to the mark and the extinctions measured at 520 nm in the photometer.

To determine the concentration of Fe^{2+} ions in the sample, a defined sample volume is treated in the same way. If the pH of the solution is not between 3 and 9, it is buffered with a little sodium acetate. A blank is analysed at the same time as the sample.

Interfering Factors

Turbid solutions should be filtered as rapidly as possible before the determination.

Calculation of Results

The content of Fe^{2+} ions is determined from the calibration curve, taking the blank value into consideration.

6.1.13 Flocculation Test (Jar Test)

Colloids or finely suspended particles in crude water for drinking water preparation can interfere with certain treatment processes, such as sedimentation or filtration. Through the addition of flocculating agents, mainly iron or aluminium salts, the forces stabilising the dispersed particles (usually electronegative ones) can be overcome. On stirring microflocculates are first formed and from them, with reduced energy input, macroflocculates. The latter can be readily separated. To improve the flocculation process and the properties of the flocculates additional flocculating aids, such as polyacrylamide, are sometimes added.

To design a flocculation plant preliminary experiments are necessary, as the pH, acidic capacity, ionic strength, nature and quantity of the flocculating agent and flocculation aid, the energy input and the residence time need to be determined to ensure efficient flocculation. The standardised flocculation test, or jar test, is therefore important for planning and optimising water and waste water treatment.

Area of Application → water

Apparatus

Multiple stirrer system with at least six stirrers, which can be adjusted continuously between 0 and 100 min^{-1}

Graduated beakers, volume 1 to 2 l, depending on the type of instrument

Reagents and Solutions

Aluminium sulfate solution,	1.95 g aluminium sulfate $(Al_2(SO_4)_3 \cdot 18H_2O)$ or
$w\ (Al_2(SO_4)_3) = 1\%$:	1.74 g $(Al_2(SO_4)_3 \cdot 14H2O)$ are made up to 100 ml with water.
Iron sulfate solution,	1.83 g iron sulfate $(FeSO_4 \cdot 7H_2O)$ or
$w\ (FeSO_4\ or\ Fe_2(SO_4)_3) = 1\%$:	1.41 g $(Fe_2(SO_4)_3 \cdot 9H_2O)$ are made up to 100 ml with water.
Calcium hydroxide suspension,	calcium hydroxide is treated with CO_2-free (boiled) water.
$w\ (Ca(OH)_2) = 1\%$:	
Polyelectrolyte solutions, 0.05%:	quantities depending on manufacturer's instructions
Sulfuric acid, $w\ (H_2SO_4) = 1\%$	

Measurement

After placing the crude water samples in the beakers, the flocculating agent to be tested is added rapidly in increasing concentrations, e.g. 1 ml of the aluminium sulfate solution (= 10 mg/l $Al_2(SO_4)_3$) in the first beaker, 2 ml in the second etc. The stirrer is immersed in the mixture and the latter stirred for 1 min at 60 to 100 min^{-1} (stirrer speed to be laid down exactly for each test series), and then for 20 min at 30 min^{-1}. The appearance of the first visible flocculation should be recorded after the stirring phase, as should the sedimentation process within 20 min, the colour, turbidity and pH of the supernatant liquid. Ca. 100 ml of the supernatant are filtered through a filter paper and the colour and turbidity of the filtrate are determined.

If, in addition, the nature and optimal concentration of flocculation aids are to be determined, the most favourable conditions for the flocculation experiment (possibly with the quantity of flocculating agent somewhat less than the optimum) are repeated with increasing concentrations of the flocculation aid to be tested. Concentrations of 1 mg/l are seldom exceeded. A beaker with flocculating agent but no flocculation aid is tested for comparison.

The pH sometimes lies outside the range of 6 to 8 required for flocculation with aluminium salts, particularly with crude water which is not easily buffered, or it decreases due to hydrolysis of the flocculating agent. After placing the same optimal concentrations of flocculating agent/flocculating aid in the beakers, different pH values of, for example 5.0 to 8.0, are set up by adding calcium hydroxide suspension or sulfuric acid.

Supplementary Information and Calculation of Results

The procedure described above is only an example. Parameters such as pH, alkalinity, stirring speed and time, and quantities of flocculating agent and flocculation aid should be varied in a number of test series to find the optimum flocculating conditions with the lowest input of chemicals. The number of individual experiments should be limited systematically using a factorial design plan (see References, e.g. Doerffel). The advice below should be followed in assessing the results of flocculation tests and their application in water treatment plants:

– The water temperature in the test should correspond to that of the crude water.
– The alkalinity of the sample can be kept at the same level through the addition of 0.40 mg/l $Ca(OH)_2$ per mg/l $Al_2(SO_4)_3$.
– The first assessment of the flocculation is qualitative, e.g. poor, satisfactory, good, outstanding.

- The average residence times of the water to be treated in the flocculation and sedimentation chambers of the treatment works can be used as orientation values for laboratory tests.
- Sometimes different suspensions require different flocculation conditions. For mixtures of suspensions compromises between two sets of conditions are sometimes necessary.
- On applying laboratory results to the water treatment plant, those test parameters are first tried out which remove turbidity at the lowest cost and consumption of chemicals.
- Scaling up the results of laboratory tests to the scale of a large plant is not always problem free.

6.1.14 Fluoride

Fluorides can occur in waste water from the aluminium, ceramics, glass, enamel and semiconductor manufacturing industries. Increased fluoride concentrations can be present in some ground water as a result of the geology of the area. Its use as drinking water can lead to fluorosis (damage to the calcium carbonate balance in living organisms).

For water samples with low matrix influences fluoride determination can usually be carried out directly using an ion-sensitive electrode (exception: complexed fluoride). In all other cases alkaline digestion followed by distillation is necessary. A simple photometric determination process has also been described.

Areas of Application ➔ water, waste water

a) Fluoride Measurement with Ion-sensitive Electrode

Apparatus
mV-meter (resolution 0.1 mV) with fluoride electrode
pH meter with electrode
Magnetic stirrer
Thermostat equipment, temperature range 20 to 30 °C
Nickel basin, ca. 0.5 l
Nickel crucible, ca. 50 ml
Steam distillation equipment (Seel apparatus)

Reagents and Solutions

TISAB buffer solution (total ionic strength adjustment buffer):	300 g trisodium citrate ($C_6H_5O_6Na_3$) are dissolved in water. 22 g 1,2-cyclohexylenedinitrilotetraacetic acid ($C_{14}H_{22}N_2O_8 \cdot H_2O$) and 60 g sodium chloride are added and the solution made up to 1 l. It should be stored in plastic bottles.
Standard fluoride solution, β (F^-) = 1000 mg/l:	2.210 g sodium fluoride (dried for 1 h at 120 °C) are made up to 1 l with water. The solution is kept in a plastic bottle.

Sulfuric acid, w (H_2SO_4) = 72%
Phosphoric acid, w (H_3PO_4) = 87%
Solid sodium hydroxide

Calibration and Measurement

Direct Determination

25 ml TISAB buffer and 25 ml sample are mixed. The pH must sometimes be adjusted to 5.8 with hydrochloric acid or sodium hydroxide, taking the dilution into account. The solution is tempered at 25 ± 0.5 °C.

The fluoride electrode is immersed in the stirred solution. If the voltage does not change by more than 0.5 mV within 5 min, the stirrer is switched off and the reading taken. A series of fluoride standards (0.2 to 10 mg/l) is treated in the same way as the sample. The results are calculated using a calibration curve.

Inorganically Bound Total Fluoride

Pretreatment:

500 ml sample are placed in a nickel basin and the pH adjusted to 11–12 with sodium hydroxide. After evaporation to ca. 30 ml, the sample is transferred to a nickel crucible and evaporated carefully to dryness. The residue is covered with 2 g sodium hydroxide and the crucible heated to 400 to 500 °C. The melt is then dissolved in water.

Distillation:

A maximum of 50 ml of the pretreated solution is placed in a distillation flask, which is then connected to the rest of the steam distillation apparatus. 60 ml sulfuric acid and 10 ml phosphoric acid are added via a dropping funnel. The condenser exit leads into a 500 ml volumetric flask containing 20 ml sodium hydroxide, c (NaOH) = 1 mol/l. The flask is heated to 155 °C. After boiling begins water is passed through the solution and ca. 450 ml of solution are distilled over.

The contents of the volumetric flask are neutralised with hydrochloric acid using methyl red as an indicator. The volume is then made up to the mark with water. The fluoride concentration is measured as under the direct determination.

Interfering Factors

To improve equilibration the fluoride electrode is placed for 1 h before the measurement in an aqueous solution with a fluoride concentration of 0.2 mg/l. In the measurement the relationship between the voltage and the logarithm of the fluoride concentration in the sample is linear in the range 0.2 to 2000 mg/l.

In the case of waste water samples nonspecific matrix influences or interference by certain cations are eliminated through distillation.

b) Simple Photometric Fluoride Determination

This method is suitable for clear and colourless water samples.

Apparatus

Ca. 10 Nessler tubes, 100 ml

Burette or dropping pipette

Reagents and Solutions

Hydrochloric acid/sulfuric acid mixture:	100 ml hydrochloric acid, w (HCl) = 36%, and 33 ml conc. sulfuric acid are carefully added successively with stirring to 800 ml water.
Zirconium oxychloride reagent:	0.3 g zirconium oxychloride ($ZrOCl_2 \cdot 8H_2O$) are dissolved in ca. 50 ml water in a 1 l volumetric flask. A solution of 0.07 g alizarin red S (sodium alizarin sulfonate) in 50 ml water are added and the solution made up to 1 l with the hydrochloric acid/sulfuric acid mixture. This solution is ready for use after 1 h and can be kept for ca. 6 months.

Sodium arsenite solution:	0.5 g sodium arsenite ($NaAsO_2$) are dissolved in 100 ml water.
Standard fluoride solution β (F$^-$) = 10 mg/l:	the solution is prepared by diluting the standard solution β (F$^-$) = 1000 mg/l described in a).

Calibration and Measurement

Different portions of between 0 and 12 ml of the standard fluoride solution are placed in seven Nessler tubes. Water is added up to the mark so that the tubes contain fluoride solutions with concentrations of 0 to 1.2 mg/l. 100, 50, 25, and 12.5 ml of the water sample are placed in four further Nessler tubes which are also filled up to the mark. With chlorinated drinking water 1 drop of sodium arsenite solution is added (sufficient for 1 mg/l chlorine in the sample). After thorough shaking and adjusting all the tubes to the same temperature ($\pm$ 2 °C), 5 ml zirconium oxychloride reagent is added rapidly to each tube, shaking repeated and the tube allowed to stand for 60 min. The colour intensities of the yellow to pink sample solutions (and their diluted forms) are compared with those of the series of standards.

Interfering Factors

Water contents which exceed the concentrations (mg/l) given below can interfere with the fluoride determination:

chloride	2000
sulfate	300
phosphate	5
iron	2
aluminium	0.25
acidic capacity	400 (as $CaCO_3$) $\approx$ 4 mol/l.

6.1.15   Dissolved and Undissolved Substances

Dissolved and undissolved substances in water are separated by filtration. Filterable substances are defined as those undissolved substances which can be separated by a medium-coarse filter and are weighed after drying for 2 h at 105 °C.

The evaporation residue includes inorganic and organic substances which are involatile at temperatures of up to 105 °C.

The ash residue includes those substances which are weighed after heating for 1 h at red heat (600 to 650 °C). The loss on ignition is the difference between the evaporation and ash residues. It arises from organic materials and through the decomposition of inorganic salts, such as nitrates, carbonates and hydrogen carbonates.

The total salt content of water can be determined by passing the sample through a cation exchanger followed by acid titration.

Areas of Application → water, waste water, soil (dry residue and ash)

a) Filterable Substances

Apparatus
Glass funnel or G2 sintered glass funnel
Medium-coarse filter paper

Measurement
For filtration sufficient sample is used to ensure that at least 10 mg solids are collected. They are washed with water, dried at 105 °C (paper filter in a weighing glass: 2 h, sintered glass funnel: ca. 20 min) and cooled in a desiccator before weighing.

Calculation of Results
The concentration β of the filterable material (mg/l) is calculated from:

$$\beta \text{ (filterable material)} = m \cdot 1000/V$$

m weight of residue, mg
V sample volume, ml

b) Evaporation Residue

Apparatus
Water bath or laboratory radiation heater
Porcelain crucibles (15 cm diameter)

Measurement
The porcelain crucible is dried at 105 °C and weighed. At least 100 ml of the water sample filtered through a medium-coarse filter paper are then evaporated to dryness in the crucible. The evaporation can be carried out in stages. The sample is then dried at 105 °C to constant weight. For soil samples, 5 to 10 g of air-dried fine soil are dried at 105 °C in a porcelain crucible to constant weight.

Calculation of Results

The concentration β of the evaporation residue (mg/l) is calculated according to:

$$\beta \text{ (evaporation residue)} = m \cdot 1000/V$$

m weight of residue, mg
V sample volume, ml

c) Ash Residue

Apparatus

Porcelain crucible (15 cm diameter)
Muffle oven

Reagents and Solutions

Ammonium nitrate solution: 1 g ammonium nitrate is dissolved in 100 ml water.

Measurement

The porcelain crucible is heated to 600 °C and weighed. The evaporation residue is placed in the crucible and heated for 1 h at 600 °C. If black residues are formed as a result of heating, they are moistened after cooling with a few drops of ammonium nitrate solution, dried carefully and heated for a further 10 min at 600 °C. The basin is weighed after cooling in a desiccator.

For soil samples the dry residue is heated in a porcelain crucible to constant weight at 600 °C.

Calculation of Results

The concentrations β_1 and β_2 of the ash residue and the loss on ignition (mg/l) are calculated according to:

$$\beta_1 \text{ (ash residue)} = m_1 \cdot 1000/V$$
$$\beta_2 \text{ (loss on ignition)} = (m-m_1) \cdot 1000/V$$

m_1 weight of ash residue, mg
m weight of evaporation residue, mg
V sample volume, ml

d) Total Salt Content

Apparatus

Glass column with tap (ca. 1 cm internal diameter, ca. 20 cm long)
Glass wool
pH paper
Titration equipment

Reagents and Solutions

Strongly acidic cation exchanger

Sodium hydroxide, c (NaOH) = 0.1 mol/l

Methyl red indicator solution

Sulfuric acid, w (H$_2$SO$_4$) = 12%

Measurement

Ca. 5 ml cation exchanger resin are allowed to swell for 30 min in water and carefully charged to the column. Conditioning and regeneration are carried out by passing ca. 20 ml sulfuric acid, w (H$_2$SO$_4$) = 12%, through the column at a rate of 1 ml/min. The column is then washed with water until the pH of the column is the same as that of the water.

The measurement is performed by passing ca. 50 ml of the filtered water sample through the column at a rate of 1 ml/min. After washing the column twice with 10 ml demineralised water, the percolate is titrated against sodium hydroxide, c (NaOH) = 0.1 mol/l, with methyl red as the indicator. The resin capacity must be taken into consideration in the measurement. The saturation point is ca. 0.9 mg equivalents per ml of resin.

Calculation of Results

The result can be calculated in terms of the concentration β of the monovalent Na$^+$ ion (in mg/l):

$$\beta\,(\text{Na}^+) = V_1 \cdot 2.23 \cdot 1000 \text{ mg} \cdot \text{ml}^{-1}/V$$

V_1 volume of sodium hydroxide consumed, c (NaOH) = 0.1 mol/l, ml

V sample volume, ml

1 mg Na$^+$ corresponds to 2.62 mg NaCl.

6.1.16   Total Organic Carbon (TOC) and Dissolved Organic Carbon (DOC)

Total organic carbon (TOC) is defined as the organically bonded carbon in a water sample, while dissolved organic carbon (DOC) is the dissolved part present in the sample. In addition water usually contains carbonates, i.e. inorganically bonded carbon, which is either removed before the determination or determined separately and subtracted from the total carbon determined.

The basis of the TOC and DOC determination is the complete oxidation of the carbon of organic substances in water to carbon dioxide. This is carried out thermally by incineration at high temperatures or by UV irradiation at normal temperature. The CO$_2$ formed is determined by IR spectrometry in most commercial instruments. As this is not an absolute determination, the process needs to be calibrated. For samples which contain undissolved substances, the TOC is best determined by incineration. After filtration through a 0.45 μm membrane filter the DOC can also be determined on the same sample. Besides the determination with a specially designed instrument, a cuvette test has been described which is particularly suitable for plant analysis.

Areas of Application → water, waste water

Apparatus
TOC/DOC analyser (gases according to manufacturer's instructions)
Ultrasound apparatus for homogenising water samples

Reagents and Solutions

Standard phthalate solutions β_1 (Pht) = 1000 mg/l, β_2 (Pht) = 100 mg/l:	2.125 g potassium hydrogen phthalate ($C_8H_5O_4K$) are dissolved in water and made up to the mark in a 1 l volumetric flask. The solution can be kept for ca. 4 weeks at 4 °C. 100 ml of this solution are made up to 1 l with water in a volumetric flask. The solution can be kept for ca. 1 week at 4 °C.
Sodium carbonate solution β (Na_2CO_3) = 1000 mg/l:	4.404 g sodium carbonate (dried for 1 h at 250 °C) are dissolved in water and made up to 1 l in a volumetric flask. The solution can be kept for ca. 4 weeks at 4 °C.

Sample Preparation
To determine the TOC the sample is homogenised for 10 min in the ultrasound bath. To determine the DOC the sample is filtered through a previously well washed 0.45 μm membrane filter.

Calibration and Measurement
To calibrate the instrument with the solutions described above, a five point calibration within the expected measurement range of the samples is recommended. The required quantities of standard solution are used for this. The measurement is carried out according to the manufacturer's instructions, predominantly within the linear portion of the calibration curve.

In some instruments a catalyst is used to improve the incineration of the sample. As the equilibrium vapour pressure is only reached above 908 °C (at normal pressure), the incineration temperature should be higher. Volatile organic compounds are incompletely determined by this method.

Measurement Using a TOC Cuvette Test (e.g. Dr Lange)
The principle of the cuvette test is based on the wet chemical oxidation of the TOC. The inorganic carbon (TIC) is coverted into CO_2 by acidification. This diffuses through a semipermeable membrane into an indicator solution during digestion at 100 °C. The colour change of the indicator is determined photometrically.

In the expulsion method the TIC is removed quantitatively before the determination of TOC. This procedure is particularly useful when there is more TIC than TOC. The sample is acidified, stirred for 5 min and placed in the cuvette together with the digestion reagent. After connecting this to the indicator cuvette, both are heated for 2 h at 100 °C in a thermostat. The coloration of the indicator cuvette is measured using a photometer.

The difference method is better for samples which have more TOC than TIC or which contain volatile organic compounds. In the first cuvette the sum of TOC and TIC is determined and in the second, only the TIC. The TOC is the difference between the two measurements.

6.1.17   Humic Substances

Humic substances are formed in natural soils and bodies of water as higher molecular weight intermediate and final products of biochemical degradations. Concentrations of between 1 and 5 mg/l are mostly found in surface water (water appears colourless), but in a few cases, as in black water rivers in tropical areas, concentrations of up to 50 mg/l are found (yellow coloration). The concentration in ground water is usually below 1 mg/l. Ca. 80% of the total quantity of soluble humic substances in most surface water consists of fulvic acid and the remainder of humic acids. In sewage plants humic substances are formed in the biological treatment stage, from which the run-off is usually yellow.

In the preparation of drinking water and water for industrial use humic substances can cause interference in the following ways:

– In the filtration through activated charcoal higher molecular weight humic acids occupy the active areas of the charcoal.
– The rate of flocculation is decreased in crude water.
– Humic substances form undesired haloforms after chlorination of drinking water.

In general the depth of colour represents a measure of the degree of decomposition of humic substances. The extinction increases as the wavelength decreases, so the quotient of the extinctions at 468 and 644 nm is often taken as a measure of the concentration of humic substances ($Q_{4/6}$ value). The measured extinctions only represent relative values.

Area of Application → water

Apparatus
Centrifuge
Spectrophotometer or fixed filter photometer with 468 and 644 nm filters
Apparatus for membrane filtration, 0.45 μm membrane filter
pH meter with electrode

Reagents and Solutions
Hydrochloric acid, c (HCl) = 1 mol/l
Sodium hydroxide, c (NaOH) = 0.1 mol/l
Standard humic acid (e.g. Merck)

Calibration and Measurement
The water sample is filtered through the membrane and is treated with hydrochloric acid, c (HCl) = 1 mol/l, to pH 1 with stirring. The solution is left overnight, the supernatant liquid carefully decanted, and the residue consisting of humic acids placed in a centrifuge tube. The residue is purified by rinsing several times with water acidified to pH 1 until the supernatant remains clear after centrifuging. The residue is dried in a weighing glass at 105 °C and weighed.

The extinctions of the supernatant after precipitation by acid (fulvic acids) and the humic acids from the precipitate, which have been redissolved in sodium hydroxide, can also be measured photometrically at 468 and 644 nm. Commercially available standard humic acid is used as the control solution.

6.1.18   Potassium

The concentration of potassium in natural water seldom exceeds 20 mg/l, whereas in some waste water and especially seepage from waste dumps, very high concentrations can be found, even exceeding the sodium level.

Areas of Application ➔ water, waste water, soil

Apparatus
Flame photometer with 768 nm filter

Reagents and Solutions

Standard potassium solutions, β_1 (K$^+$) = 100 mg/l, β_2 (K$^+$) = 10 mg/l:	1.907 g potassium chloride (dried at 105 °C) are made up to 1 l with water. This solution contains β (K$^+$) = 1000 mg/l. Standard solutions for which β_1 (K$^+$) = 100 mg/l and β_2 (K$^+$) = 10 mg/l are prepared from it by dilution. The solutions are stored in plastic bottles.

Sample Preparation
Water samples are filtered before measurement to prevent blockage of the suction equipment of the photometer.
   To determine the available potassium in soil, 1 part air-dried soil and 10 parts water are shaken for 15 min and the potassium in the filtered extract is determined.

Calibration and Measurement
A calibration curve is drawn by measuring the emission intensities of standards at 768 nm in the desired range (e.g. 0 to 1 mg/l or 0 to 10 mg/l). The sample and a blank are then analysed.

Interfering Factors
The presence of high concentrations of sulfate, chloride or bicarbonate can cause interference. The standard addition technique can extensively eliminate these types of problem. Known quantities of potassium are added to the sample with an unknown potassium concentration. The intensity measured is plotted against the quantity of potassium added and the curve is extrapolated to the point of intersection with the negative part of the abscissa. The value read off at this point gives the potassium concentration of the unknown solution (Fig. 34).

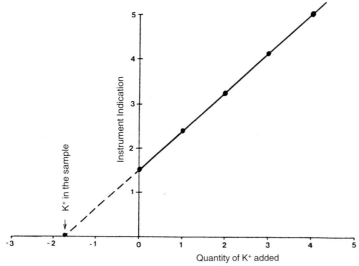

Fig. 34: Standard addition technique

Comparison of these curves with the curves plotted using standard solutions gives information on possible interference or matrix effects.

Calculation of Results
The results are calculated from the calibration curve.

6.1.19   Silicic Acid

Silicon occurs in all rocks and sediments. Silicon compounds, such as silicic acid, can be dissolved from these materials by weathering processes and thus reach the water cycle. Silicic acid can be found in dissolved, colloidal or suspended form. Its concentration in natural waters is frequently between 0 and 20 mg/l, while in strongly mineralised ground water higher concentrations can be found. In water for industrial use silicic acid is undesirable because it can form deposits in pipes or boilers.

Silicic acid ($Si(OH)_4$) is only stable for a short period at pH 3.2. Otherwise it is transformed by dehydration into orthosilicic acid ($H_6Si_2O_7$), then into polysilicic acid and finally into metasilicic acid ($H_2SiO_3)_n$. Its solubility decreases with increasing molecular size and degree of dehydration. $Si(OH)_4$ is readily soluble in water, while SiO_2, as the last member of the chain of silicic acids, is insoluble.

The following method describes the determination of the reactive 'soluble' silicic acid after reaction with molybdate.

Area of Application → water

Apparatus
Spectrophotometer or fixed filter photometer with 812 and 650 nm filters

Reagents and Solutions

Ammonium molybdate solution:	10 g ammonium molybdate $((NH_4)_6Mo_7O_{24} \cdot 4H_2O)$ are made up to 100 ml with warm water. After filtration and adjustment of the pH to 7, the solution is stored in a plastic bottle.
Oxalic acid solution:	10 g oxalic acid $(C_2H_2O_4 \cdot 2H_2O)$ are made up to 100 ml with water.
Sodium carbonate solution:	25 g anhydrous sodium carbonate are dissolved in 1 l water.
Standard silicic acid solution, $\beta\,(SiO_2) = 10$ mg/l:	either commercially available standard solutions are used, or, if these are unobtainable, standards are prepared as follows: 1 g silicon dioxide (SiO_2) is heated to 1100 °C in a platinum crucible for ca. 1 h and after cooling 5 g anhydrous sodium carbonate are added. The mixture is heated at red heat until it melts. When gas is no longer given off, heating is continued at light red heat for a further 10 min. After cooling, the melt is dissolved in water and the solution made up to 1 l. The solution is stored in a plastic bottle. It contains $\beta\,(SiO_2) = 1000$ mg/l. 10 ml are removed from this solution and made up to 1 l.

Hydrochloric acid, $w\,(HCl) = 20\%$

Sample Preparation
Before determination of the dissolved silicic acid the sample is filtered through a 0.45 μm membrane filter. If colloidal silicic acid is also to be determined, the sample must be digested. 100 ml sample and 20 ml sodium carbonate solution are placed in a platinum crucible and carefully evaporated down to 80 ml. This solution is transferred to a 100 ml volumetric flask, 5 ml hydrochloric acid, $w\,(HCl) = 20\%$, are added and the solution made up to the mark with water.

Calibration and Measurement
Aliquots of between 1 and 10 ml standard silicic acid solution are taken and made up to 50 ml. The solutions are analysed together with 50 ml of sample by treating them with 1 ml hydrochloric acid, $w\,(HCl) = 20\%$, and 2 ml ammonium molybdate solution. After mixing, 1.5 ml oxalic acid solution are added and mixing repeated. After 5 min the measurement is carried out in the photometer at 812 or 650 nm. Owing to the lower sensitivity at 650 nm, 0 to 30 ml of the standard should be used to construct the calibration curve. A blank is treated in the same way as the sample.

Interfering Factors
Phosphates, iron and sulfides can cause interference. Interference by phosphate is reduced by the addition of oxalic acid. Colour interference can be compensated for by photometric comparison measurements.

Calculation of Results

The concentration of silicic acid is calculated from the calibration curve.

6.1.20   Kjeldahl Nitrogen

Kjeldahl nitrogen is the sum of ammonium nitrogen and those organic nitrogen compounds which can be converted into ammonium under Kjeldahl reaction conditions. The proportion of organic nitrogen can be calculated by subtraction of the ammonium content from that of Kjeldahl nitrogen.

Low concentrations of soluble organic nitrogen compounds can be present in natural waters through the degradation of organic matter. Further enzymatic conversion to ammonium is possible. Waste water generally contains higher concentrations of organic nitrogen compounds.

Areas of Application ➔ water, waste water, soil

Apparatus
Kjeldahl flasks, 350 ml
Distillation apparatus with 1 l flasks, Liebig condenser, heater

Reagents and Solutions

Sodium hydroxide solution:	400 g sodium hydroxide are dissolved in 1 l water.
Reaction mixture:	5 g selenium, 5 g anhydrous copper sulfate, and 250 g anhydrous sodium sulfate are mixed in a mortar and stored under dry conditions.
Phenolphthalein solution:	1 g phenolphthalein is dissolved in 100 ml ethanol and 100 ml water added.
Mixed indicator solution:	a) 30 mg methyl red are dissolved in 100 ml ethanol; b) 100 mg methylene blue are dissolved in 100 ml water; 100 ml solution a) are mixed with 15 ml solution b).

Conc. sulfuric acid
Sulfuric acid, c (H_2SO_4) = 0.025 mol/l

Measurement
A 100 ml water sample is placed in the Kjeldahl flask and 1 g of reaction mixture and 10 ml ethanol are added. For soils, 1 to 5 g air-dried fine soil are used. After shaking, 10 ml conc. sulfuric acid are added and the mixture heated to boiling until the colour is light green and no more black particles are present. Boiling is then continued for 30 min. Nitrite and nitrate are removed by this procedure. After cooling and diluting with water to a total volume of ca. 300 ml, the solution is transferred to the 1 l flask of the distillation apparatus and the Kjeldahl flask rinsed twice. A few drops of phenolphthalein solution are added followed by sodium hydroxide until the contents of the flask are red.

The flask is connected to the distillation apparatus and ca. 200 ml distilled over. The exit tube from the condenser should dip into the absorption medium. The ammonium concentration is then determined titrimetrically or photometrically, depending on the original nitrogen content. For nitrogen concentrations up to $\beta\,(N) = 10$ mg/l the photometric method is preferred, while the titrimetric method is usually used for higher concentrations.

Titrimetric determination:

The distillate is collected in a 250 ml volumetric flask containing 50 ml water. 100 ml of this solution are mixed with 3 drops of the mixed indicator solution and titrated against sulfuric acid, c $(H_2SO_4) = 0.025$ mol/l, until a colour change from violet to green is observed. A blank water sample is titrated in the same way.

Photometric determination:

50 ml of the distillate are made up to 250 ml and determined as described under 'ammonium' (see Section 6.1.2).

Interfering Factors

Various aromatic and heterocyclic nitrogen compounds are not completely determined by this method.

Calculation of Results

Titrimetric determination:

The concentration β of Kjeldahl-N in mg/l is given by:

$$\beta\,(\text{Kjeldahl-N}) = a \cdot b \cdot 700 \text{ mg} \cdot \text{ml}^{-1}/c \cdot d$$

a consumption of c $(H_2SO_4) = 0.025$ mol/l, ml
b aliquot of distillate taken, ml
c sample volume, ml
d total volume of distillate, ml

Photometric determination:

The results are calculated from the calibration curve.

6.1.21   Hydrocarbons

Contamination by aliphatic hydrocarbons can occur in surface water and waste water and, after certain pollution incidents, also in ground water. This class of substance consists essentially of lipophilic, straight-chain or branched hydrocarbons with saturated or unsaturated bonds. Examples include mineral oils and tallows, petroleum spirit, some industrial solvents, plant and animal oils, fats and waxes. The substances are either dissolved or are present as emulsions or as a separate phase. Surfactants generally promote emulsion formation. In crude water destined for drinking water hydrocarbons impair the taste and odour even at very low concentrations and must therefore be removed. During waste water disposal fatty acids can be liberated from organic oils and fats through saponification and can lead to corrosion of concrete pipes.

According to DIN 38409 a distinction has conventionally been made between the following:

a) Involatile lipophilic substances with a boiling point > 250 °C:
Through extraction of the water sample all substances which can be extracted by the chosen solvent are removed and are determined gravimetrically after evaporation of the solvent. With substances with boiling points < 250 °C lower values are obtained depending on the method. Results obtained using different extraction solvents should not be compared with one another.

b) All substances which can be extracted into 1,1,2-trichlorotrifluoroethane ($C_2Cl_3F_3$) and which remain after removal of polar materials:
The hydrocarbons are separated by extraction and the co-extracted polar substances are removed by the polar absorption agent aluminium oxide. For the determination of hydrocarbons by IR spectrometry the characteristic absorbtions of the CH_3-, CH_2- and CH-groups are used. The specific measure of spectroscopic absorption depends on the composition of the hydrocarbon mixture. Aromatics are only determined to a small extent.

c) All substances which can be extracted into 1,1,2-trichlorotrifluoroethane and which can be separated directly by gravity:
The method is principally used to determine hydrocarbons present in waste water as a separate phase and the dimensioning and monitoring of separators for low density liquids at concentrations of 5 mg/l to 10 g/l.

   The procedures described below generally do not give any information on defined component classes or individual compounds.

a) Gravimetric Method

The extraction is preferably carried out using 1,1,2-trichlorotrifluoroethane. The (nonflammable) solvent can be replaced by hexane or petroleum ether, but their high flammability must be taken into consideration. A differentiation between saponifiable oils and fats (plant and animal) and nonsaponifiable mineral components may be made.

Areas of Application ➜ water, waste water

Apparatus
Separating funnels, 1 l
Water bath
Centrifuge
250 ml flasks with ground glass stoppers
Reflux condenser

Reagents and Solutions

1,1,2-Trichlorotrifluoroethane ($C_2Cl_3F_3$),	or alternatively n-hexane or petroleum ether
Alcoholic potassium hydroxide, c (KOH, alc.) = 0.1 mol/l:	ethanol is used as the alcohol.
Ethanol	

Sample Preparation

Samples should be collected in glass bottles, previously cleaned with solvents.

Measurement

The extraction is performed in the sample bottle by shaking with 25 ml solvent for 1 min. After phase separation the aqueous phase is pipetted into a second bottle containing 25 ml solvent and shaken again for 1 min. The aqueous phase is then discarded. The extracts are transferred to a 1 l separating funnel and the bottle rinsed with solvent. The remaining aqueous phase (or solvent phase in the case of 1,1,2-trichlorotrifluoroethane) is run out. The solvent is passed through a filter paper, containing a little anhydrous sodium sulfate for drying, into a weighed, constant weight glass crucible (or preferably a platinum crucible). If an emulsion is formed, the mixture is centrifuged. The solvent is removed on a water bath at ca. 80 °C. The crucible is dried for 5 min at 105 °C, allowed to cool in a desiccator, and the weight of the residue determined.

Should soaps be included in the analysis, the sample must be first acidified to pH 1–2 so that the fatty acids liberated can be extracted together with oils and fats. The soaps can also be determined separately from the oils and fats by acidifying the already extracted sample and reacidifying.

To separate saponifiable oils and fats from nonsaponifiable components, the residue in the crucible is taken up in ethanol, transferred to a 250 ml flask with a ground glass stopper and 50 ml alcoholic potassium hydroxide are added. After refluxing for 60 min, the contents of the flask are transferred to a separating funnel, the flask rinsed with 10 to 50 ml solvent and the oils extracted by shaking in the separating funnel. The phases are separated and the alcoholic potassium hydroxide re-extracted after addition of solvent. The combined solvent extracts are filtered through a filter paper into a crucible and treated as described above. The residue remaining after evaporation of the solvent extracts contains the nonsaponifiable components of the oils and fats.

Interfering Factors

The determination is relatively nonspecific as different components, such as emulsifiers, waxes or detergents, can also be partially or completely extracted.

Emulsions can significantly impair the determination, but they can be broken down by the addition of sodium sulfate, by acidification or centrifugation.

Strongly contaminated samples or sludges are first evaporated down on a water bath. The residues are transferred quantitatively to an extraction thimble and extracted for several hours in a Soxhlet apparatus. The extract is then treated as described above.

Calculation of Results

The concentration β of the extractable oils and fats, in mg/l, is calculated as follows:

$$\beta \text{ (extractable oils and fats)} = a/V$$

a weight of extraction residue, mg
V sample volume, l

b) IR Spectrometric Method

Hydrocarbons which are extractable with 1,1,2-trichlorotrifluoroethane can be determined quantitatively using IR spectrometry from the extinctions of the CH-vibrations at 2958 cm^{-1} (CH$_3$-band), 2924 cm^{-1} (CH$_2$-band) and 3030 cm^{-1} (aromatic CH-band). Co-extracted polar hydrocarbons, such as vegetable oils, higher alcohols, ketones or esters, can be removed quantitatively or at least partially by column chromatography using aluminium oxide.

Areas of Application ➔ water, waste water

Apparatus
IR spectrophotometer with quartz cuvettes
Stirrer (e.g. Ultraturrax), shaker or magnetic stirrer
250 ml separating funnel
Chromatography column according to DIN 38409, part 18: length 10 cm, diameter 2 cm with glass frit

Reagents and Solutions
Aluminium oxide: 100–200 μm (70–150 mesh ASTM), neutral, activity grade I
1,1,2-Trichlorotrifluoroethane
Anhydrous sodium sulfate
Reference substance: e.g. squalane (C$_{30}$H$_{62}$)

Calibration and Measurement
After collection, 1 l water sample (a smaller volume at higher hydrocarbon concentrations) is placed together with 25 ml 1,1,2-trichlorotrifluoroethane in a wide-necked bottle with a conical shoulder and shaken in the laboratory for 10 min (30 sec is sufficient with a high-speed stirrer). The mixture can be acidified with sulfuric acid to pH 1 beforehand to improve phase separation. After phase separation, most of the upper aqueous layer is removed and the remainder is transferred to a separating funnel. The sample bottle is rinsed with a little solvent. If a stable emulsion is formed, the organic phase is centrifuged. After renewed phase separation in the separating funnel, the organic phase is passed through a fluted filter paper containing ca. 10 g sodium sulfate on to the chromatography column charged with 8 g aluminium oxide. The liquid is run through into a 50 ml volumetric flask. The solvent used to rinse the containers is also passed through the filter and column into the volumetric flask. After filling up to the mark, the liquid is ready for measurement. The quantity of hydrocarbons contained in the extract should not exceed 10 mg; otherwise it should be diluted.

   The IR measurement is performed according to the manufacturer's instructions in a quartz cuvette between 3200 and 2700 cm^{-1}. Calibration can be carried out using a standard mixture of the same hydrocarbons as in the sample if they are known. If the composition of the extracted substances is not known, squalane is frequently used as a standard.

Calculation of Results

The results are calculated according to DIN 38409 part 18. As the transmission T is often determined with IR spectrometers, this is converted into the corresponding extinction using the formula $E = -\log T$.

gasoline mixtures (mg/l):

$$pG = \frac{(1300 \cdot V_{TE})}{d \cdot V_s} \cdot (E_1/c_1 + E_2/c_2 + E_3/c_3)$$

Hydrocarbon mixtures similar to mineral oil (mg/l):

$$pM = \frac{(1400 \cdot V_{TE})}{d \cdot V_s} \cdot (E_1/c_1 + E_2/c_2)$$

pG   sample concentration (gasoline), mg/l
pM   sample concentration (mineral oil hydrocarbons), mg/l
V_{TE} volume of extraction agent used, ml
d path length of the absorbing solution, cm
V_s sample volume, ml
E_1 extinction of the CH_3-band at 3.38 μm (v = 2958 cm^{-1})
c_1 group extinction coefficient of the CH_3-band determined for 8.3 ml/mg · cm for a number of mineral oil products
E_2 extinction of the CH_2-band at 3.42 μm (v = 2924 cm^{-1})
c_2 group extinction coefficient of the CH_2-band determined for 5.4 ml/mg · cm
E_3 extinction of the CH-band at 3.30 μm (v = 3030 cm^{-1})
c_3 group extinction coefficient of the CH band determined for 0.9 ml/mg · cm

Example: gasoline

$$pG = \frac{1300 \cdot 25\ ml}{1000\ ml \cdot 1\ cm} \cdot (0.5\ mg \cdot cm/8.3\ ml + 0.52\ mg \cdot cm/5.4\ ml + 0.25\ mg \cdot cm/0.9\ ml)$$

$$= 14\ mg/l$$

Interfering Factors

The procedure involves a sum determination, which generally does not give any information on defined component classes or individual compounds. In individual cases the method is used as a screening method to establish whether relevant quantities of hydrocarbons are present in the samples at all. In addition the following should be noted:

–   During the work-up losses of low-boiling components are possible.
–   The polar or aromatic compounds contained in the extract are more or less quantitatively separated from the rest of the components by adsorption on to aluminium oxide and are therefore

not determined in the IR measurement. However, because some aromatic components have pronounced lipophilic character, the extract purification described can lead to false interpretations.
–   If the aluminium oxide chromatography step is omitted, the interpretation of the results is problematic because it is very difficult to estimate which substances affect the result.
–   The following information should be given with the results: extraction solvent, nature and duration of the extraction, separation, nature of the reference substance (e.g. squalane or hydrocarbon mixture).

c) Directly Separable Lipophilic Low Density Materials

Low density materials, such as gasoline and fuel, diesel, and lubricating oils, are present in directly separable, emulsified or dissolved form, depending on their degree of dispersion. Directly separable lipophilic, low density materials are defined as the portion which separates immediately, i.e. without further physico-chemical pretreatment, as a consequence of its density and degree of dispersion. The densities and water solubilities of some mineral oil products are listed in Table 25.

Table 25: Density and water solubility of some mineral oil products

Substance	Density (15 °C) (g/cm^3)	Solubility in water (20 °C) (mg/l)
Gasoline	0.71–0.79	100–150
Benzene	0.88	1800
Diesel fuel	0.82–0.86	~ 20
Gear oil	0.89–0.94	~ 1
Extra light fuel oil	0.82–0.86	5–20
Heavy fuel oil	0.90–1.05	< 1
Kerosene	0.75–0.84	5–10
Naphtha	0.70–0.76	~ 15
Paraffin oil	0.88–0.90	~ 1
Crude oil	~ 0.80	5–10
Toluene	0.87	580
o-Xylene	0.87	180

Area of Application ➔ waste water

Apparatus
Five to ten 1 l separating funnels with taps and glass stoppers, graduated to 50 ml in the lower tapered section
Laboratory balance, maximum load 3000 ± 1 g

Sample Collection
5 to 10 individual samples are collected successively directly with the separating funnels. For the separation of low density materials it should be ensured that the sampling point represents the total population. In the case of rivers and waste water channels sampling is carried out at points where the turbulence is such that it is possible to obtain a representative sample.

Measurement
The empty separating funnels are weighed without their stoppers before sampling. Immediately after collection of ca. 1 l samples, the filled dried vessels are reweighed and the volume calculated from the weight difference and the density.

After allowing 15 ± 1 min for separation (with the stopper in), liquid is allowed to run out of each separating funnel at a uniform rate until the phase boundary still remains visible. The volume of the organic phase is then read off. For most tests the precision of this direct reading is adequate. For smaller organic phase volumes or for high-precision requirements, 40 ml 1,1,2-trichlorotrifluoroethane are added to the contents of each separating funnel. The funnel is shaken and, after phase separation, the lower organic phase is filtered through 10 g anhydrous sodium sulfate into a 50 ml volumetric flask. The separating funnel is rinsed with 10 ml solvent.

The directly separable lipophilic low density materials can be determined by the methods described under a) or b). The aqueous phases run off from the separating funnels can also be analysed for their dissolved and emulsified involatile lipophilic components and hydrocarbons using methods a) or b).

6.1.22   Copper

In natural unaffected waters the concentration of copper does not exceed a few µg/l. In bodies of water contaminated with copper, concentrations of 0.1 to 0.2 mg/l can be toxic to lower water organisms. Higher concentrations in drinking water are generally a result of corrosion of copper pipes. Concentrations of up to 3 mg/l are not uncommon after a long standing time. A relatively simple photometric method is described below. However interference problems are greater than with the atomic absorption spectrometry method requiring more complex equipment (see Section 6.1.30).

Area of Application → water

Apparatus
Spectrophotometer or fixed filter photometer with 440 nm filter
250 ml separating funnel

Reagents and Solutions

DDTC solution:	1 g sodium diethyldithiocarbaminate ($C_5H_{10}NS_2Na$) is dissolved in 100 ml water. The solution is only stable for a few days.

Standard copper solution,	1 g metallic copper is dissolved in 10 ml nitric acid,
β (Cu) = 10 mg/l:	w (HNO$_3$) = 30%, and made up to 1 l with water. 10 ml of this solution are taken and made up to 1 l with water.

Citric acid solution, w (C$_6$H$_8$O$_7$) = 20%

Sulfuric acid, w (H$_2$SO$_4$) = 35%

Ammonium chloride solution, w (NH$_4$Cl) = 20%

Chloroform

Calibration and Measurement

Several aliquots of between 0 and 30 ml of the standard copper solution (corresponding to m (Cu) = 0 to 0.3 mg) are taken and treated in the same way as the water sample.

100 ml sample (if necessary made up to 100 ml) are placed in a separating funnel and 1 ml citric acid, w (C$_6$H$_8$O$_7$) = 20%, 0.5 ml sulfuric acid, w (H$_2$SO$_4$) = 35%, 0.5 ml ammonium chloride solution, w (NH$_4$Cl) = 20%, and 10 ml chloroform are added. The contents of the vessel are shaken for a few min and then the organic phase is discarded. 2 ml DDTC solution and 25 ml chloroform are added and shaking continued for 3 min. The organic phase is then filtered and measured at 440 nm in the photometer.

Interfering Factors

Higher concentrations of zinc and heavy metals can cause interference. The copper complex extracted into chloroform is light-sensitive so the measurement must be carried out immediately after extraction.

Calculation of Results

The copper content is calculated from the calibration curve.

6.1.23   Manganese

Manganese is often found in clean surface waters at concentrations of several tenths of a mg/l. However, levels exceeding 1 mg/l can occur under anaerobic conditions. As is the case with iron (together with which this element is often found) the presence of manganese in mains water is generally undesirable because it can lead to incrustation by oxides. In addition, even small quantities can be detrimental to taste. The removal of manganese from crude water destined for drinking is even more difficult than the removal of iron. Normally the water is aerated and then passed through a gravel filter which has a manganese oxide surface already formed on it.

The analysis of manganese can be carried out using the reaction with formaldoxime or the oxidation to permanganate by peroxodisulfate. The former method is preferred for relatively clean water, and the latter for polluted and discoloured water.

Areas of Application → water, waste water

a) Determination with Formaldoxime

Apparatus
Spectrophotometer or fixed filter photometer with 480 nm filter

Reagents and Solutions

Formaldoxime solution:	4 g hydroxylammonium chloride ($NH_2OH \cdot HCl$) and 0.8 g paraformaldehyde $(CHOH)_x$ are made up to 100 ml with water.
Ferrous ammonium sulfate solution:	0.14 g ferrous ammonium sulfate $((NH_4)_2Fe(SO_4)_2 \cdot 6H_2O)$ are made up to 100 ml with water and 1 ml conc. sulfuric acid.
EDTA solution:	40 g disodium EDTA $(C_{10}H_{14}O_8N_2Na_2 \cdot 2H_2O)$ are dissolved in 100 ml water.
Hydroxylammonium chloride solution:	10 g hydroxlammonium chloride ($NH_2OH \cdot HCl$) are made up to 100 ml with water.
Ammonia solution:	75 ml ammonia solution, w (NH_4OH) = 25%, are mixed with 25 ml water.
Standard manganese solutions, β_1 (Mn) = 100 mg/l, β_2 (Mn) = 10 mg/l:	308 mg manganese sulfate ($MnSO_4 \cdot H_2O$) are dissolved in water and made up to 1 l after addition of 3 ml conc. sulfuric acid. 100 ml of this solution are made up to 1 l with water.

Sample Preparation
Calcium and magnesium concentrations of more than 300 mg/l lead to artificially high values, so in these cases dilution is necessary.

Suspended particles are removed by centrifugation prior to the photometric determination.

Measurement
The following reagents are added sequentially to 50 ml of the water sample with shaking: 5 ml formaldoxime solution, 5 ml ferrous ammonium sulfate solution, 5 ml ammonia solution, and then after 5 min 5 ml EDTA solution and 5 ml hydroxylammonium chloride solution.

After a minimum of 1 h the extinction of the solution is measured at 480 nm. A blank is treated in an identical manner.

Interfering Factors
Interference by ferrous ions is prevented by addition of EDTA and hydroxylammonium chloride.

Phosphate concentrations of more than 10 mg/l together with the presence of calcium ions can result in values which are too low.

Discoloured samples are analysed by the second method described below.

Calculation of Results
The results are calculated from the calibration curve.

b) Determination as Permanganate

Apparatus
Spectrophotometer or fixed filter photometer with 525 nm filter

Reagents and Solutions

Reaction solution:	7.5 g mercuric sulfate are dissolved in 40 ml nitric acid, w (HNO_3) = 65%, and 20 ml water. 20 ml orthophosphoric acid, w (H_3PO_4) = 85%, and 3.5 mg silver sulfate are added and the solution made up to 100 ml with water.
Standard manganese solution:	as described under a)
Ammonium peroxodisulfate ($NH_4)_2S_2O_8$	

Sample Preparation
The sample should be acidified immediately after collection to prevent precipitation of insoluble manganese compounds. As permanganate ions react with reducing agents, these must be removed before the determination. Suspended particles are filtered off. Organic materials (from 60 mg/l $KMnO_4$ consumption) are removed by oxidation with nitric acid. This involves evaporating 100 ml of the sample containing 1ml conc. sulfuric acid and 1 ml nitric acid, w (HNO_3) = 65%, until white fumes appear. In cases of brown coloration some water is added followed by repeated small quantities of nitric acid. The residue is taken up in dilute nitric acid and the solution made up to 100 ml with water.

Calibration and Measurement
5 ml reaction solution and 1 g solid ammonium peroxodisulfate are added to 90 ml of the sample (after pretreatment if necessary). After boiling for 1 min and cooling under running water, the sample is made up to 100 ml with water. The extinction is measured at 525 nm in the photometer. A blank is treated in the same way.

Interfering Factors
Interference caused by turbidity and larger quantities of dissolved organic substances is prevented as described above.

Chloride ions up to ca. 1000 mg/l are masked by addition of mercury ions. At higher concentrations the sample must be pretreated as described above.

Calculation of results
The results are calculated from the calibration curve.

6.1.24 Sodium

Sodium is one of the major components of many natural waters with concentrations of up to several hundred mg/l. High to very high concentrations are found in domestic waste water, some industrial waste water, seepage water from waste dumps and in sea water. Detectable concentrations are also found in rain water, depending on the distance from the coast. In soils in arid regions the content of sodium ions plays an important role in the problem of oversalting.

Areas of Application ➜ water, waste water, soil

Apparatus
Flame photometer with 589 nm filter

Reagents and Solutions

Standard sodium solutions, β_1 (Na) = 100 mg/l, β_2 (Na) = 10 mg/l:

2.542 g sodium chloride (dried at 105 °C) are made up to 1 l with water. This solution contains β (Na) = 1000 mg/l. Standard solutions with β_1 (Na) = 100 mg/l and β_2 (Na) = 10 mg/l are prepared by dilution. The solutions are stored in plastic bottles.

Sample Preparation
Turbid samples are filtered before measurement to avoid blockage of the suction equipment in the apparatus.

For determination of the available sodium in soil, 1 part air-dried soil is shaken with 10 parts distilled water and the sodium in the filtered extract is determined.

Calibration and Measurement
A calibration curve is constructed by measuring the emission intensities of standards at 589 nm in the desired range (e.g. 0 to 1 mg/l or 0 to 10 mg/l). The sample and a blank are then measured. Alternatively the standard addition technique may be used to compensate for possible matrix effects (see Fig. 34, Section 6.1.18). Here various known quantities of sodium are added to the sample of unknown concentration. The measured intensities are plotted against the quantities added and the curve extrapolated until it cuts the negative part of the abscissa. The point of intersection gives the sodium concentration in the sample.

Comparison of this curve with that obtained using pure standard solutions can give information on possible interference or matrix effects.

Interfering Factors
The presence of sulfate, chloride or bicarbonate in concentrations of more than 1000 mg/l can cause interference. The standard addition method can almost entirely eliminate such problems.

Calculation of Results
The results are obtained by reference to the calibration curve.

6.1.25   Nitrate

Nitrate is found in many natural waters at concentrations of between 1 and 10 mg/l. Higher concentrations often indicate the effects of nitrogen-containing fertilisers since the nitrate ion is only poorly adsorbed in soil and easily reaches the ground water. High nitrate concentrations are found in outflows from modern sewage plants because ammonium nitrogen is partially or completely nitrified microbiologically. However nitrate concentrations in untreated waste water are low. The concentration of nitrate is an important parameter for the assessment of self-purification properties of bodies of water and the nutrient balance in surface water and soil. The photometric determination with sodium salicylate is described below.

Areas of Application → water, waste water, soil

Apparatus
Spectrophotometer or fixed filter photometer with 420 nm filter
Water bath or sand bath
Porcelain crucibles (8 to 10 cm diameter)

Reagents and Solutions

Sodium salicylate solution:	0.5 g sodium salicylate ($C_7H_5O_3Na$) are made up to 100 ml with water. The solution must be freshly prepared daily.
Alkaline tartrate solution:	400 g sodium hydroxide and 60 g sodium potassium tartrate ($C_4H_4O_6KNa \cdot 4H_2O$) are made up to 1 l with water.
Standard nitrate solutions,	1.6307 g potassium nitrate are made up to 1 l with water.
β_1 (NO_3^-) = 1000 mg/l,	10 ml of this solution are taken and made up to 1 l.
β_2 (NO_3^-) = 10 mg/l:	
Conc. sulfuric acid	

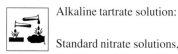

Sample Preparation
The sample should be analysed as rapidly as possible, especially in the case of waste water. Strong discoloration can normally be removed by precipitation with aluminium hydroxide (see Section 6.1.2).

Calibration and Measurement
The calibration is carried out by placing volumes of the standard solutions corresponding to 0.01–0.5 mg/l nitrate in porcelain crucibles. 2 ml sodium salicylate solution are added and the mixture carefully evaporated on a water bath. The residue is dried for 2 h at ca. 100 °C. After cooling 2 ml conc. sulfuric acid are added and the mixture left for 10 min. 15 ml water and 15 ml alkaline tartrate solution are then added. The solution is finally transferred to a 100 ml volumetric flask, the solution made up to the mark, and the nitrate concentration determined photometrically at 420 nm within 10 min. The values obtained are used to construct a calibration curve.

For the analysis of the water sample the volume is chosen according to the expected nitrate concentration. For concentrations of more than 100 mg/l the sample must be diluted, otherwise the analysis is the same as for the standard solution. A blank is measured in the same way.

For the determination of nitrate in soil an extract with demineralised water (1:10) is shaken for 2 h. The extract is analysed as described above. Turbid solutions need to be filtered before measurement.

Interfering Factors

Interference caused by the presence of more than 200 mg/l chloride can be prevented by dilution. Nitrite causes interference at concentrations above 1 mg/l and can be eliminated by addition of 50 mg ammonium sulfate to the sample and evaporation to dryness before continuing with the analysis.

Calculation of Results

The nitrate concentration is determined from the calibration curve and the blank value.

6.1.26   Nitrite

Nitrite is generally not present in unpolluted waters or only in very small concentrations. It is formed as an intermediate in the microbiological nitrification of ammonium to nitrate if the bacterium *Nitrobacter* is inhibited by a lack of oxygen or bactericidal action of pollutants, for example. Nitrite is toxic to many organisms at concentrations well below 1 mg/l. A threshold value of 0.1 mg/l is laid down in the ordinance on drinking water.

Areas of Application → water, waste water

Apparatus

Spectrophotometer or fixed filter photometer with 540 nm filter

Reagents and Solutions

Reagent solution:	0.5 g sulfanilamide ($C_6H_8N_2O_2S$) and 0.05 g N-(1-naphthyl)-ethylenediamine dihydrochloride ($C_{12}H_{16}Cl_2N_2$) are dissolved in 25 ml water and 10.5 ml hydrochloric acid, w (HCl) = 36%. 13.6 g sodium acetate ($C_2H_3O_2Na \cdot 3H_2O$) are added and the solution made up to 50 ml. The solution is stable for several months.
Standard nitrite solutions, β_1 (NO_2^-) = 1000 mg/l, β_2 (NO_2^-) = 1 mg/l:	1.50 g sodium nitrite (dried for 1 h at 105 °C) are made up to 1 l with water. This solution is stable for ca. 1 month at 4 °C. 1 ml of this solution is made up to 1 l with water. It should be freshly prepared daily.

Sample Preparation

The analysis should be carried out within a few hours of sample collection. In all cases the sample must be kept cool until examination. Discoloration and colloidal turbidity may be removed by aluminium hydroxide flocculation (see Section 6.1.2). At pH > 10 or a base capacity exceeding 6 mmol/l, the pH value is adjusted to 6 with dilute hydrochloric acid.

Calibration and Measurement
For the calibration aliquots of 1 to 25 ml of the dilute standard nitrite solution are pipetted into 50 ml volumetric flasks, diluted to ca. 40 ml, and treated with 2 ml reagent solution. After making up to the mark, mixing and allowing to stand for 15 min, the extinctions are measured at 540 nm.

Then 40 ml of the sample are analysed (or at higher nitrite concentrations a smaller quantity is taken and made up to ca. 40 ml). The pH of the solution should be between 1.5 and 2. The analysis is carried out as described above. A blank is treated in the same way as the sample.

Interfering Factors
Nitrogen oxides present in the laboratory air can falsify the results. Higher concentrations of strong oxidising and reducing agents (e.g. active chlorine, sulfite) can cause interference.

Calculation of Results
The nitrite concentration is calculated by reference to the calibration curve and the blank value.

6.1.27   Phenol Index

Aromatic hydrocarbons with hydroxyl groups on the aromatic ring are known as phenols. They are designated as mono-, di-, or polyhydric, depending on the number of OH groups. They are found in small quantities in natural waters because, as components of plants, they are liberated during degradation and humification processes. Certain types of untreated industrial waste water (e.g. that from coking plants or from the production of synthetic resins or disinfectants) contain considerable phenol concentrations.

The toxicity of phenols depends on the nature and arrangement of functional groups in the molecule. Chlorination of drinking water can lead to the formation of chlorophenols from humic substances, which give the water an unpleasant taste and odour.

For the purposes of analysis phenols are divided into steam-volatile and non-steam-volatile compounds. The volatility decreases in the following order: cresols, xylenols, phenol, naphthol, catechol, hydroquinone, and other polyhydric phenols. The steam volatility is pH-dependent; it is higher the lower the pH.

The conventional procedure for determination of the phenol index includes all phenols which undergo the coupling reactions described below.

Areas of Application → water, waste water

Sample Preparation
Immediately after collection the sample is adjusted to pH > 12 with sodium hydroxide or to a pH of 3 to 4 with hydrochloric acid and is kept cold.

a) Phenol Index without Distillation ('Total Phenols')

Apparatus
Spectrophotometer or fixed filter photometer with 530 nm filter
250 ml separating funnel

Reagents and Solutions

p-Nitroaniline solution:	700 mg p-nitroaniline ($C_6H_6N_2O_2$) are dissolved in 150 ml hydrochloric acid, c (HCl) = 1mol/l, and the solution made up to 1 l with water.
Saturated sodium nitrite solution:	ca. 85.5 g sodium nitrite are dissolved in 100 ml water at 25 °C.
Standard phenol solutions,	1 g phenol is made up to 1 l with water. By taking aliquots
β_1 (Ph) = 10 mg/l,	from this solution the standards required are freshly prepared
β_2 (Ph) = 1 mg/l	daily.

β_3 (Ph) = 0.1 mg/l:
Sodium carbonate solution, c (Na_2CO_3) = 1 mol/l
Sodium hydroxide, w (NaOH) = 30%
n-Butanol

Calibration and Measurement
At least 50 ml of the water sample are placed in the separating funnel with 30 ml sodium carbonate solution and the pH adjusted to 11.5 if necessary. 20 ml p-nitroaniline solution are treated with
a few drops of sodium nitrite solution. This mixture is added to the sample and the resulting mixture allowed to stand for 20 min. The dye formed is extracted by shaking with 50 ml butanol and
the aqueous phase is removed after 10 min. The extinction of the butanol extract is measured at 530
nm against a parallel running blank. For calibration various solutions are prepared from the standard phenol solution, which cover the concentration range of the measurement.

b) Phenol Index after Distillation ('Steam-distillable Phenols')

Apparatus
Spectrophotometer or fixed filter photometer with 460 and 510 nm filters
1 l separating funnel
Distillation apparatus with 1 l round-bottomed flask, Liebig condenser, heater, 500 ml measuring
cylinder as receiver

Reagents and Solutions

Amino-antipyrine solution:	2 g 4-amino-2,3-dimethyl-1-phenyl-3-pyrazoline-5-one ($C_{11}H_{13}N_3O$) are dissolved in water and made up to 100 ml.
Buffer solution (pH 10.0):	34 g ammonium chloride and 200 g sodium potassium tartrate ($C_4H_4O_6KNa \cdot 4H_2O$) are dissolved in 700 ml water, treated with 150 ml ammonia solution, w (NH_4OH) = 25%, and made up to 1 l with water.

| Buffer solution (pH 4.0): | 151 g disodium hydrogen phosphate ($Na_2HPO_4 \cdot 2H_2O$) and 142 g citric acid ($C_6H_8O_7 \cdot H_2O$) are made up to 1 l with water. |
| Potassium hexacyano-ferrate(III) solution: | 8 g potassium hexacyanoferrate(III) ($K_3[Fe(CN)_6]$) are made up to 100 ml with water. The solution must be stored away from light and is stable for ca. 1 week. |

Standard phenol solutions as described under a)
Anhydrous sodium sulfate
Copper sulfate ($CuSO_4 \cdot 5H_2O$)
Ferric sulfate ($Fe_2(SO_4)_3 \cdot 9H_2O$)
Trichloromethane (chloroform)

Calibration and Measurement

500 ml of the sample (possibly diluted) are placed in the distillation flask and 0.5 g copper sulfate added. The solution is shaken several times over a period of 10 min and 50 ml buffer solution (pH 4) are added. If necessary, the pH value is adjusted to 4. 400 ml are then distilled over into the measuring cylinder, which is topped up to 500 ml with water, and the contents transferred, together with 20 ml buffer solution (pH 10), to the separating funnel. 3 ml amino-antipyrine solution are added and, after shaking for a short time, 3 ml potassium hexacyanoferrate(III) solution. Shaking is then continued. After leaving for 30 to 60 min (protected from light) the dye formed is extracted by shaking with 25 ml chloroform for 5 min. The organic phase is filtered through 5 g anhydrous sodium sulfate into a 25 ml volumetric flask. After rinsing the separating funnel the solution is made up to the mark with chloroform. The colour intensity of the solution is measured at 460 nm in the photometer against chloroform as a reference. At higher concentrations of phenolic substances the chloroform extraction step can be dispensed with. In such cases, a smaller sample volume is employed and the photometric measurement performed at 510 nm.

Interfering Factors

Oxidising agents present in the sample (e.g. chlorine or chlorine dioxide) interfere but their effects can be minimised by the addition of ascorbic acid. In the presence of reducing agents a small amount of ferric sulfate is added before the distillation. Should the sample have its own colour, a blank without addition of amino-antipyrine is treated similarly and its value subtracted from the result.

Calculation of Results

The phenol concentration is obtained by reference to the calibration curve and the blank value.

6.1.28 Phosphorus Compounds

Natural unaffected waters mostly contain total phosphorus at concentrations of less than 0.1 mg/l. Contaminated surface water, however, contains higher phosphorus concentrations caused by the discharge of waste water and the washing out of fertiliser residues used on fields. Too much phosphorus in still bodies of water often leads to eutrophication, lowering of the oxygen concentration and problems in water treatment. Phosphorus compounds are adsorbed on to soil particles to such an extent that the danger of seepage into the deeper layers or even into the ground water is relatively slight.

Phosphorus compounds are determined in water, waste water or soil in various forms: total phosphorus, soluble orthophosphate, hydrolysable phosphate and organically bonded phosphate. The method of analysis depends on the forms present.

Orthophosphates are those which are to be determined without prior hydrolysis. Condensed phosphates are converted into orthophosphates by hydrolysis of the sample under weakly acidic conditions. During this process, conversion of certain organic phosphate components into inorganic phosphates is unavoidable. Only that portion of the total phosphorus which is digested under strong oxidising conditions is referred to as 'organic phosphorus'.

For problems concerning the eutrophication of water systems, the determination of dissolved orthophosphates and the total phosphorus is recommended. At higher concentrations of both orthophosphate and the proportion of total phosphorus which can be determined as orthophosphate, conditions favourable for eutrophication are always found whereas at higher total phosphorus and nondetectable amounts of orthophosphate, propagation of organisms hardly occurs.

The determination of the sum of orthophosphates and condensed phosphates is necessary in the cases of drinking water and water for industrial use (treatment with phosphates to reduce corrosion).

Areas of Application ➜ water, waste water, soil

a) Orthophosphate

Apparatus
Spectrophotometer or fixed filter photometer with 880 or 700 nm filter

Reagents and Solutions

Ammonium molybdate solution:	40 g ammonium molybdate ($(NH_4)_6Mo_7O_{24} \cdot 4H_2O$) are made up to 1 l with water.
Ascorbic acid solution:	2.6 g L-(+)-ascorbic acid ($C_6H_8O_6$) are dissolved in 150 ml water. This solution must be freshly prepared.
Potassium antimony(III) oxytartrate solution:	2.7 g potassium antimony(III) oxytartrate ($K(SbO)C_4H_4O_6 \cdot 2H_2O$) are dissolved in water and made up to 1 l.
Reaction mixture:	250 ml sulfuric acid, w (H_2SO_4) = 25%, 75 ml ammonium molybdate solution and 150 ml ascorbic acid are mixed. 25 ml potassium antimony(III) oxytartrate solution are added and the solution mixed again. The solution should be freshly prepared daily and stored in the refrigerator.
Standard phosphate solution, β (PO_4^{3-}) = 1 mg/l:	139.8 mg potassium dihydrogen phosphate (KH_2PO_4) (dried at 105 °C) are made up to 1 l with water. 10 ml of this solution are taken and made up to 1 l.

Sample Preparation

The sample is passed through a 0.45 μm membrane filter as soon as possible after collection. The first 10 ml are discarded.

Calibration and Measurement

Volumes of the standard phosphate solution corresponding to between 0 and 0.04 mg are placed in five 50 ml volumetric flasks and are made up to 40 ml with water. These solutions and maximum 40 ml of the sample are then treated with 8 ml reaction mixture and made up to 50 ml with water. After mixing, the solutions are left for 10 min and measured at 880 or 700 nm in the photometer.

The blank correction for colour and turbidity is made by adding to the water sample 8 ml of a reaction mixture containing only sulfuric acid and potassium antimony(III) oxytartrate in the amounts given above.

Interfering Factors

Silicic acid at a concentration exceeding 5 mg/l can appear as a high phosphate concentration. Chromate has the opposite effect and can be reduced by addition of 1 ml ascorbic acid. Sulfides at concentrations of more than 2 mg/l can be removed by addition of several mg potassium permanganate. After shaking, the excess reagent is reduced by addition of 1 ml ascorbic acid solution.

Calculation of Results

The orthophosphate concentration is obtained from the calibration curve.

Conversion: 1 mg PO_4^{3-} = 0.326 mg P

b) Hydrolysable Phosphate (Total Inorganic Phosphate)

Reagents and Solutions

In addition to the solutions described under a):

Phenolphthalein solution: 1 g phenolphthalein is dissolved in 100 ml ethanol and 100 ml water are added.

Conc. sulfuric acid

Sodium hydroxide, w (NaOH) = 20%

Measurement

Up to 40 ml of the sample are treated with 1 ml conc. sulfuric acid. After boiling for 5 min, cooling and neutralisation using phenolphthalein solution and sodium hydroxide, w (NaOH) = 20%, the mixture is diluted to ca. 40 ml. The further procedure is described under a). A blank is treated in the same way as the sample.

Calculation of Results

The concentration of hydrolysable phosphate is determined from a calibration curve. The result gives the sum of orthophosphate and condensed phosphates. The hydrolysable phosphate is obtained by subtracting the orthophosphate from this value.

c) Total Phosphorus

Reagents and Solutions
In addition to the solutions described under a) and b):

Potassium peroxodisulfate solution:	5 g potassium peroxodisulfate ($K_2S_2O_8$) are dissolved in 100 ml water. The solution should be freshly prepared daily.

Measurement
100 ml sample (or a smaller quantity diluted to 100 ml) are treated with 0.5 ml conc. sulfuric acid to give a pH < 1. 15 ml potassium peroxodisulfate solution are then added and the solution boiled gently for 30 min (up to 90 min for organic phosphorus compounds which are difficult to digest.). After cooling, 1 drop of phenolphthalein solution is added followed by enough sodium hydroxide solution to give a pink colour. The solution is made up to 100 ml with water and the phosphate content determined as described under a). The standard solution and the blank are treated in the same way as the sample.

Calculation of Results
The concentration of total phosphorus is determined from the calibration curve. The concentration of organically bonded phosphorus is obtained by subtraction of the quantity of hydrolysable phosphate from the total phosphorus.

6.1.29   Sludge Volume and Sludge Index

Sludge volume and sludge index are two important parameters for the characterisation of sewage sludge, predominantly activated sludge. The sludge volume is defined as the volume in ml of the sludge-water mixture in the activating basin of a sewage plant, after a sedimentation time of 30 min. The sludge index is the volume occupied by 1 g of dry substance of the activated sludge.

Areas of Application → sewage sludge

Apparatus
1 l glass or transparent plastic measuring cylinder, diameter 6–7 cm

Measurement
The sample should be taken quickly from the activating basin, i.e. without allowing separation. It is placed in the measuring cylinder and the latter filled up to the mark. The vessel is allowed to stand for 30 min without shaking. The sludge volume is then read at the level of the boundary between the sludge and the supernatant. The determination should be repeated if the sludge volume exceeds 250 ml. In this case the sample should be diluted with the supernatant water of the same sample in the ratio 1 + 1 or 1 + 2. The sludge volume determined should then be multiplied by the factor 2 or 3.

The dry residue of the sample should be determined parallel to this measurement (see Section 6.1.15).

Interfering Factors

Higher sludge volumes can cause interference with the sedimentation process. In these cases the sample is diluted. Gas bubbles can also hinder sedimentation. At temperature differences of $> 3\,°C$ between the sample in the measuring cylinder and the surrounding air the measuring cylinder is placed in a bucket filled with water at the same temperature as the sample to avoid convection currents.

Calculation of Results

The sludge index (ml/g) is calculated according to:

$$I = S/T$$

I sludge index of the sludge – water mixture, ml/g
S sludge volume of the sample, ml/l
T dry residue of the sample, g/l

The sludge index of a readily flocculating activated sludge lies between 50 and 100 ml/g, while values of > 200 ml/g indicate bulking sludge.

6.1.30   Heavy Metals by Atomic Absorption Spectrometry (AAS)

Heavy metals are usually present in water and soil in very small concentrations as a result of contact with certain minerals. Through human activity the concentrations can increase considerably, giving rise to toxic effects on organisms. Regulations for the protection of bodies of water and soil therefore always contain threshold values for heavy metals.

The methods of determination for six selected heavy metals are described in the following DIN standards (flame AAS and graphite tube AAS):

– lead: DIN 38406, part 6,
– cadmium: DIN 38406, part 19,
– chromium: DIN 38406, part 10,
– copper: DIN 38406, part 7,
– nickel: DIN 38406, part 11,
– zinc: DIN 38406, part 8.

For the measurement of other elements the standard procedures and other literature must be referred to (e.g. Rump/Scholz, 1995).

For many water, waste water and soil samples the detection and determination limits of flame AAS are sufficient for the determination of elements. Only in a few cases, to improve the detectability limits, is the use of graphite tube AAS necessary. Manufacturers and users usually give different instructions on the detection and determination limits. This is partially due to differences of opinion regarding matrices, interfering factors and optimisation of instruments.

Apparatus
Atomic absorption spectrometer with background compensation and suitable radiation source
Graphite tube system with automatic injection unit and control device for the heating programme
Supplies of the following pure gases: air, acetylene and dinitrogen oxide (N_2O)
Microwave digestion unit

Reagents and Solutions

Standard solutions:	commercially available solutions are frequently used to prepare standard solutions. Stock solutions for the elements Cd, Cu, Ni, Pb and Zn can be prepared by dissolving 1000 mg metal in 10 ml conc. nitric acid and making up the solution to 1 l. To prepare the chromium stock solution 2.825 g $K_2Cr_2O_7$ (dried for 2 h at 105 °C) are made up to 1 l with water and 5 ml nitric acid. Suitable standards are obtained by diluting the stock solution together with 10 ml/l nitric acid.

Sample Preparation
At higher organic contamination 20 to 50 ml sample are treated with 5 ml 65% nitric acid and heated in a pressurised vessel in the microwave digestion unit for 2 min at 400 W and then for 7 min at 300 W (repeat if digestion is not complete). After cooling, 0.5 ml 30% hydrogen peroxide are added and the digestion continued under the conditions described above (for other digestions see Section 6.3.1.4).

Calibration and Measurement
Flame AAS
The instrument parameters for the elements to be measured are adjusted according to the manufacturer's instructions.
Suitable wavelengths are:

lead:	217.0 nm (with interference 283.3 is better), air/acetylene flame,
cadmium:	228.8 nm, air/acetylene flame,
chromium:	357.9 nm, dinitrogen oxide/acetylene flame, addition of lanthanum chloride solution,
copper:	324.7 or 327.4 nm, at higher concentrations 249.2 or 244.2 nm, air/acetylene flame,
nickel:	232.0 nm, at higher concentrations 341.5 nm, air/acetylene flame,
zinc:	213.8 nm, at higher concentrations 307.6 nm, air/acetylene flame.

Graphite Tube AAS
The instrument parameters for the elements to be measured are adjusted according to the manufacturer's instructions.
The wavelengths are as given above.
Additives:

Cd, Pb, Zn:	200 µg $NH_4H_2PO_4$ per sample are added as matrix modifier,
Cr, Cu, Ni:	50 µg $Mg(NO_3)_2$ per sample are added as matrix modifier.

The calibration is carried out with the appropriate standard solutions. Water with 1 ml 65% nitric acid per 100 ml is used as the blank. The intensity of the spectral line is measured for the reference and blank solutions. The calibration curve it then drawn up (this is usually performed by the instrument's computer). The element concentrations in the sample are then measured in the same way.

With only slightly contaminated samples the concentrations can be calculated from the calibration curve. For higher contamination the standard addition procedure (spiking procudure) is used. If the element concentration in the sample solution exceeds that in the standard solution, it must be diluted with the blank.

Interfering Factors

Interference in the measurement with flame and graphite tube AAS is particularly likely to occur with samples with a complex matrix. If there is spectral interference alternative spectral lines can be used. Instructions for this can be found in the manufacturer's handbook. For samples with a complex or less well known matrix, the standard addition procedure should be used.

6.1.31   Sulfate

Sulfate ions occur in natural unaffected waters at concentrations of up to 50 mg/l, though concentrations of up to more than 1000 mg/l or more can be found in water in contact with certain geological formations (gypsum reserves, water from pyrite quarries). In anaerobic ground water aquifers the concentrations are usually low while those of hydrogen sulfide are higher as a result of sulfate reduction.

Contaminated bodies of water and waste water normally have high sulfate concentrations, usually as a result of industrial discharges, waste dumps or fertilisers.

Sulfate-containing water is calcium carbonate aggressive: a sulfate content of > 250 mg/l is slightly damaging to concrete, while one of > 600 mg/l is highly damaging. High sulfate concentrations corrode steel pipes.

High sulfate contents alter the taste of drinking water. Above 250 mg/l the water may act as a laxative.

Two methods for sulfate determination are described below: the gravimetric method, which is employed where precision is important, and the turbidimetric method, which is less precise but quicker to carry out.

Areas of Application ➔ water, waste water, soil

a) Gravimetric Method

Apparatus
Drying oven
Muffle furnace
Water bath
Quartz or platinum crucibles

Reagents and Solutions

Barium chloride solution:	10 g barium chloride ($BaCl_2 \cdot 2H_2O$) are dissolved in 90 ml water.
Methyl orange indicator:	100 mg methyl orange are made up to 100 ml with water.
Silver nitrate solution:	1 g silver nitrate is dissolved in 100 ml water together with a few drops of nitric acid.

Hydrochloric acid, w (HCl) = 20%

Sodium chloride solution, w (NaCl) = 10%

Sample Preparation

Suspended particles must be filtered off before analysing water samples or soil extracts. Silicates cause interference at concentrations of > 25 mg/l and can be removed together with organic substances as follows:

A sample volume containing no more than 50 mg/l sulfate is evaporated almost to dryness on the water bath. A few drops of hydrochloric acid, w (HCl) = 20%, and a few drops of sodium chloride solution, w (NaCl) = 10%, are added. The mixture is then evaporated to dryness so that the salt crust is in contact with the acid. After ashing at ca. 500 °C, the residue is moistened with 3 ml water and a few drops of hydrochloric acid and again evaporated to dryness. The sample is then dissolved in a little hot water and 1 ml hydrochloric acid. Ca. 50 ml hot water are added and the solution filtered. The filter residue contains insoluble silicic acid and is washed with water until no more chloride is detected on testing with silver nitrate solution. The filtrate and washings are used for the sulfate determination.

Measurement

A volume of sample containing up to 50 mg/l sulfate is placed in a beaker and made up to 200 ml if necessary. The solution is adjusted to pH 7 using the methyl orange indicator. 2 ml hydrochloric acid, w (HCl) = 20%, are added and the mixture boiled briefly. Hot barium chloride solution is then added with stirring until the precipitation appears to be complete. An excess of 3 ml is then added. Heating is continued for a further 30 min and the mixture left for at least 2 h (best overnight) before filtering through ash-free filter paper or a constant weight porcelain sintered disc (A1). The precipitate is washed with hot water until the chloride reaction is negative. The filter paper is transferred to a constant-weight porcelain crucible and carefully dried and ashed, during which the paper should not be permitted to burn in the open flame. It is then heated for 30 min at ca. 800 °C. In the case of a porcelain sintered disc it should be heated to constant weight at 300 °C and then weighed after cooling in a desiccator.

Interfering Factors

Interference by organic compounds, nitrates and silicic acid can be prevented by the sample preparation as described. Heavy metals can cause lower values to be obtained as they hinder the complete precipitation of barium sulfate.

Calculation of Results

The sulfate concentration, in mg/l, of the sample is calculated according to:

$$\beta\,(SO_4^{2-}) = a \cdot 0.4115/V$$

a   weight of barium sulfate, mg
V   sample volume, l

b) Turbidimetric Method

Apparatus

Spectrophotometer or fixed filter photometer with 420 nm filter
Magnetic stirrer

Reagents and Solutions

Barium chloride:	$(BaCl_2 \cdot 2H_2O)$, solid
Conditioning reagent:	30 ml hydrochloric acid, w (HCl) = 36%, 300 ml water, 100 ml ethanol and 75 g sodium chloride are mixed. 50 ml glycerol $(C_3H_8O_3)$ are added and the solution mixed.
Standard sulfate solution $\beta\,(SO_4^{2-}) = 100$ mg/l:	147.9 mg anhydrous sodium sulfate are made up to 1 l with water.

Sample Preparation

Turbid water samples are filtered before analysis. Samples with sulfate concentrations > 50 mg/l must be diluted.

Calibration and Measurement

Standard solutions with sulfate concentrations in the range 0 to 50 mg/l and the sample are all made up to volumes of 100 ml where necessary. They are treated with 5 ml conditioning reagent and stirred continuously with a magnetic stirrer. 0.2 to 0.3 mg solid barium chloride are added with stirring and stirring is continued for exactly 1 min. The solution is then transferred immediately to a cuvette and the absorption at 420 nm measured several times over a period of 2 to 3 min. The highest measured value is noted. A blank, consisting of a water sample without addition of barium chloride, is treated in the same way.

Calculation of Results

The concentration of sulfate ions is determined from the calibration curve.

6.1.32 Sulfide

Hydrogen sulfide and sulfides can be formed during bacterial degradation of proteins and by sulfate reduction under strongly anaerobic conditions. Water samples with measurable hydrogen sulfide concentrations have an unpleasant odour and are therefore not suitable for use as drinking water without further treatment.

The method given in DIN 38 405, part 27 can be used for the determination of sulfide in water samples. Dissolved and most undissolved sulfides are determined. For rapid testing of water, sludge and sediment samples the potentiometric sulfide determination can be used.

Areas of Application → water, waste water

a) Photometric Determination

Apparatus
Degassing apparatus for sulfide separation (Fig. 35)
pH meter with electrode
Spectrophotometer or fixed filter photometer with 665 nm filter

Reagents and Solutions

Zinc acetate solution:	20 g zinc acetate ($C_4H_6O_4Zn \cdot 2H_2O$) are dissolved in water and the solution made up to 1 l.
EDTA solution:	100 g disodium EDTA ($C_{10}H_{14}N_2O_8Na_2 \cdot 2H_2O$) are dissolved in 940 ml water.
Phthalate buffer:	80 g potassium hydrogen phthalate ($C_8H_5KO_4$) are dissolved in 920 ml water and the pH adjusted to 4.0.

Reagent solution:	2 g N,N-dimethyl-1,4-phenylenediammonium chloride are placed in a 1 l volumetric flask. 200 ml water are added, followed by 200 ml conc. sulfuric acid (care!). After cooling the volume is made up to 1 l.

Ferric ammonium sulfate solution:	50 g ferric ammonium sulfate ($NH_4Fe(SO_4)_2 \cdot 12H_2O$) are treated with 10 ml conc. sulfuric acid in a 500 ml volumetric flask. Water is added carefully up to the mark.
Sulfide solutions:	ca. 3.5 g sodium sulfide ($Na_2S \cdot xH_2O$) (x = 7–9) are dissolved in water and the volume made up to 1 l. The sulfide sulfur is determined iodometrically. The solution can be kept for 3 days. Standard solutions are prepared from this solution by dilution with water, taking the results of the iodometric titration into account.

Sample Preparation
10 ml zinc acetate solution are added to 490 ml water sample. After mixing the pH is adjusted to 8.5–9. The sample should be kept cool during transportation and analysed rapidly.

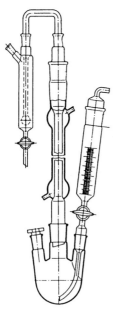

Fig. 35: Degassing apparatus for sulfide separation

Calibration and Measurement

25 ml phthalate buffer and 5 ml EDTA solution are placed in the flask and 20 ml zinc acetate solution in the absorption vessel of the degassing apparatus. A nitrogen stream of ca. 40 l/h is passed through the apparatus for 10 min. 50 ml sample is added from a dropping funnel (at > 1.5 mg/l sulfide the sample should be diluted). The dropping funnel is rinsed and nitrogen is passed through the apparatus for 60 min.

10 ml reagent solution and 1 ml ferric ammonium sulfate solution are added to the contents of the absorption vessel. The latter is filled up to the mark with water, closed, shaken and allowed to stand for 10 min. The solution is then transferred quantitatively to a 100 ml volumetric flask. The extinction of the solution is measured at 665 nm against water. The sulfide concentration in the sample is determined using a calibration curve.

A blank using water is treated in the same way as the sample.

Interfering Factors

The following ions do not cause interference in the degassing process provided that their concentrations in mg/l given below are not exceeded:

cyanide   2
sulfite   700
thiosulfate 900

b)   Determination with an Ion-selective H₂S Electrode

Apparatus
pH/ mV-meter
H_2S electrode (e.g. Ingold H_2S-245-85)

Reagents and Solutions

Standard sulfide solutions:	see above
Reference solutions L_1, L_2:	two 16.6 g portions of potassium iodide are dissolved in two standard buffer solutions of pH 4.01 (L_1) and 6.87 (L_2) respectively and each one made up to 1 l with the corresponding buffer. The solutions should be allowed to stand for 3 days before use.
Solid silver iodide	

Calibrating and Measurement
Calibration of the Electrode with Sulfide Solutions:
Increasing quantities of standard sulfide solution are added to 50 ml of a tempered buffer solution through which nitrogen has been bubbled and the potential is measured with the H_2S electrode.

Indirect Electrode Calibration:
A spatula tip of silver iodide is added to 50 ml of a tempered reference solution L_1. The solution is stirred and the potential across the electrode is determined after ca. 10 min. The electrode is then rinsed and the process repeated with solution L_2. After each measurement the electrode is conditioned in a solution of sulfide (β (S^{2-}) = 0.1 mol/l) in the appropriate buffer.

To take the measurement the electrode is dipped into the sample and the potential across the electrode is read off after ca. 10 min.

The calculation of the hydrogen sulfide concentration is performed according to the instructions given by the electrode manufacturer, taking the temperature and the pH into consideration.

6.1.33   Surfactants

Surfactants are synthetic surface-active substances, which generally consist of mixtures of isomeric or homologous individual compounds. The surfactant molecule always has one or more hydrophobic and hydrophilic groups. The former give the molecule its surface-active properties and the latter a degree of water-solubility. Depending on the nature of ion formation, one can differentiate between anionic, cationic, nonionic and amphoteric surfactants.

Surfactants are used in the home and in industry. Anionic surfactants currently form the largest part of the total production, but the proportion of nonionic and cationic surfactants is increasing.

Surfactants get into bodies of water mainly via waste water pipes or household waste dumps and can cause problems (e.g. reduction of oxygen diffusion, foaming). In many countries the use of so-called hard surfactants, i.e. not readily biodegradable materials, instead of long-chain degradable ones is no longer allowed.

A procedure for determining anionic surfactants by reaction with methylene blue is described below, as well as a rapid test for nonionic surfactants using the Dragendorff reagent. For the exact measurement of smaller concentrations and the elimination of possible matrix interference, pre-enrichment is necessary. This method and also a method for the determination of cationic surfactants can be found from the large number of handbooks (e.g. Rump/Scholz, 1995).

Areas of Application ➔ water, waste water

a) Anionic Surfactants

Apparatus
Spectrophotometer or fixed filter photometer with 650 nm filter
500 ml separating funnel

Reagents and Solutions

| Methylene blue reagent: | 30 g methylene blue are dissolved in 50 ml water and treated with 6.8 ml conc. sulfuric acid and 50 g sodium dihydrogen phosphate ($NaH_2PO_4 \cdot H_2O$). The mixture is then made up to 1 l with water. |

Wash solution: 6.8 ml conc. sulfuric acid and 50 g sodium dihydrogen phosphate are dissolved in water and the solution made up to 1 l.

Standard surfactant solutions,   1 g sodium lauryl sulfate is dissolved in water and the
β_1 (S) = 1000 mg/l,   solution made up to 1 l. The solution must be kept in the
β_2 (S) = 10 mg/l: refrigerator. 10 ml of this solution are made up to 1 l with water. The latter must be freshly prepared daily.

Trichloromethane (chloroform)
Sodium hydroxide, *c* (NaOH) = 1 mol/l

Sample Preparation
The sample must be diluted according to the expected amount of methylene blue active substance (MBAS). At concentrations of 10 to 100 mg/l MBAS, 2 ml of sample is used, at 2 to 10 mg/l MBAS, 20 ml sample, and at lower MBAS concentrations 100 to 400 ml sample. Before measurement the samples are rendered slightly alkaline by dropwise addition of sodium hydroxide, *c* (NaOH) = 1 mol/l, using phenolphthalein as indicator. The pink colour is then removed by adding a minimum quantity of dilute sulfuric acid.

Calibration and Measurement
For construction of the calibration curve various volumes of the dilute standard solution between 0 and 20 ml are placed in five separating funnels. These standard solutions and the sample itself (in another separating funnel) are diluted to 100 ml. 10 ml chloroform and 15 ml methylene blue reagent are added to each vessel and the mixtures shaken for 30 sec. After phase separation the chloroform extract is run into a second separating funnel and the water is extracted twice with 8 ml chloroform. The combined extracts are shaken in the separating funnel for 30 sec with 50 ml wash solution and then passed through glass wool into a 50 ml volumetric flask. The wash solution is extracted twice

with 10 ml chloroform and these extracts also passed through glass wool into the volumetric flask. The glass wool is rinsed with a little chloroform and the contents of the volumetric flask made up to 50 ml with the washings and more chloroform. The extinction is measured at 650 nm within 1 h.

A blank consisting of 100 ml water is treated in the same way as the sample.

Interfering Factors

The detection of surfactants can be affected by the presence of certain similar substances lacking surfactant properties, e.g. aromatic sulfonates, organic phosphates.

Interference by proteins or alkali metal salts of higher fatty acids is minimised by buffering.

Sulfide or thiosulfate interference can be prevented by addition of a few drops of hydrogen peroxide.

Chloride concentrations exceeding 1000 mg/l interfere so these samples should be diluted before the determination.

Calculation of Results

The results are calculated using the calibration curve.

b) Nonionic Surfactants

Reagents and Solutions

Bismuth salt solution:	1.7 g basic bismuth nitrate ($BiO(NO_3) \cdot H_2O$) are dissolved in 20 ml glacial acetic acid and the solution made up to 100 ml with water.
Potassium iodide solution:	40 g potassium iodide are dissolved in 100 ml water.
Barium chloride solution:	20 g barium chloride ($BaCl_2 \cdot 2H_2O$) are made up to 100 ml with water.
Reagent (ready-to-use):	the bismuth salt and potassium iodide solutions are combined, treated with 200 ml glacial acetic acid and made up to 1 l with water. 100 ml of this solution are treated with 50 ml barium chloride solution. The reagent can be kept for ca. 14 days in brown bottles.
Nonylphenol standard, β (N.) = 100 mg/l:	100 mg nonylphenol ($C_{15}H_{24}O$) are dissolved in water and the solution made up to 1 l.

Measurement

The water sample is filtered. 5 ml of the filtrate are treated with 5 ml reagent in a test tube with shaking. The presence of nonionic surfactants of the poly-alkylene oxide-type is indicated by an orange-red precipitate. Very small quantities are only indicated by turbidity. After centrifugation the precipitate can easily be seen at the bottom of the test tube. The concentration of nonionic surfactants can be estimated semiquantitatively by visual comparison with nonylphenol standard over the range 0.1 to 5 mg/l. This screening method has a detection limit of ca. 0.1 mg/l.

6.1.34   Turbidity

Unlike most ground waters surface water often contains suspended material. In still bodies of water, such as reservoirs and lakes, the major proportion of undissolved material consists of micro-organisms, whereas in rivers suspended mineral particles predominate. Turbidity which suddenly appears in ground water can be an indication of the penetration of surface or rain water into the subterranean layers with insufficient filtration during passage through the unsaturated zone. Particle sizes of 10^{-6} m to $3 \cdot 10^{-4}$ m give rise to turbidity. For mineral particles this includes the coarse and fine clay and fine sand fractions. The limit of turbidity measurement is 1000 particles per ml. Series of measurements in waterworks have shown no significant correlation between the nephelometric turbidity value and the number of particles per ml.

Turbidity is usually determined nephelometrically using optical instruments measuring scattered light. The values obtained depend on the wavelength of the incoming light and the scattering angle. Because of the differences in the nature, size and number of particles in most water samples, exact calibration using absolute turbidity values is impossible. A suspension of formazine has been used as a reference. By using the double beam method and monochromatic light at 860 nm, yellow-brown inherent coloration of the sample can be compensated for.

Areas of Application ➔ water, waste water

Apparatus
Turbidity measuring instrument, double beam if possible, (90°-scattered light, 860 nm)

Reagents and Solutions

Hydrazine sulfate solution w ($N_2H_4 \cdot H_2SO_4$) = 1%:	1 g hydrazine sulfate is made up to 100 ml with water.
Hexamethylenetetramine solution w (Hex.) = 10%:	10 g hexamethylenetetramine ($C_6H_{12}N_4$) are made up to 100 ml with water.
Stock reagent solution:	5 ml hydrazine sulfate solution and 5 ml hexamethylenetetramine solution are mixed in a 100 ml volumetric flask. After leaving to stand for 24 h at 25 ± 3 °C the solution is made up to 100 ml with water. The solution should be freshly prepared each month. It has a defined turbidity of 400 formazine nephelometric units (FNU, identical to NTU and FTU).
Standard solution:	depending on the requirement, aliquots of the stock reagent solution are diluted with water. The solutions should be freshly prepared daily.

Calibration and Measurement
The measurement is carried out according to the manufacturer's instructions. Samples with low turbidity should be shaken well before measurement. To remove bubbles of gas the measuring cuvette can be placed in an ultrasound bath for a few seconds. Depending on the instrument used, the turbidity value is either read off directly or taken from a previously drawn up calibration curve.

Calculation of Results
At values of < 10 FNU the result is rounded off to the nearest 0.1 FNU, and at values > 10 FNU to the nearest 1 FNU.

6.1.35   UV Absorption

Many of the contents of water absorb UV radiation because of their molecular structure. Above all some non-readily biodegradable compounds with delocalised electrons, such as humic substances, which are present in ground water, surface water, sewage outflows and seepage water from waste dumps, can be detected by UV and usually also determined quantitatively. Substances with single bonds do not exhibit any pronounced UV absorption. UV absorption at 254 nm has proved useful. The monochromatic light beamed into the measuring cuvette is attenuated and the transmitted light determined using a detector. The ratio of the measured absorption and the zero value for demineralised water is given by the dimensionless spectroscopic absorption coefficient.

In the investigation of sewage plant outflows and seepage water from waste dumps it has been shown that in many cases there is a correlation between the UV absorption of a sample and the chemical oxygen demand (COD) and the dissolved organic carbon (DOC). Because these types of water usually have similar compositions, UV absorption is suitable as a rapid screening method, but only after the statistical connection between the parameters has been determined by correlation and regression calculations or graphically if necessary.

Areas of Application ➔ water, waste water

Apparatus
Spectrophotometer or fixed filter photometer with 254 nm and 580 nm filters
Quartz cuvettes with 0.5, 1 and 5 cm path lengths
0.45 μm membrane filter

Sample Preparation
Before measurement the sample must be free from turbidity, as otherwise the proportion of scattered light would falsify the results. For this reason the sample is filtered through a 0.45 μm membrane filter. The first few ml of filtrate are discarded.

Another possibility for compensating for turbidity involves measuring the absorption coefficient at 254 nm and 580 nm and correcting the measurement at 254 nm to allow for the proportion of scattered light.

Measurement
The internal transmission density (previously known as extinction) of the filtered sample is measured in a quartz cuvette with a suitable path length at 254 nm. At values of > 2 the sample is diluted with water or a cuvette with a shorter path length is used so that a linear correlation between absorption and concentration is obtained.

Alternatively the internal transmission density of the unfiltered sample can be determined at 254 and 580 nm.

Calculation of Results

The spectroscopic absorption coefficient SAC_{254} (m^{-1}) is calculated from:

$$SAC_{254} = A_{254}/d$$

A   internal transmission density
d   path length of the cuvette, m

Correcting the internal transmission density of the unfiltered sample determined at 580 nm for the proportion of scattered light allows the turbidity to be compensated for.

6.1.36   Zinc

Zinc is present in natural waters up to a concentration of 50 µg/l. Higher concentrations in mains drinking water pipes are generally caused by corrosion of galvanised steel. After a long standing time concentrations of up to 5 mg/l are not unusual. Zinc contamination can occur in bodies of water as a result of waste water discharges. Concentrations of more than 0.5 mg/l are toxic to some species of fish.

The determination of zinc through the complexing reaction with dithizone is described below. It is suitable for water samples with relatively low contamination. For samples of waste water and seepage water from waste dumps the AAS procedure should be used (see Section 6.1.30).

Area of Application → water

Apparatus
Spectrophotometer or fixed filter photometer with 535 nm filter
100 ml separating funnel

Reagents and Solutions

Dithizone solution:	10 mg dithizone ($C_{13}H_{12}N_4S$) are dissolved in 1 l chloroform. The solution is stored in a brown glass bottle.
Acetate buffer solution:	a) 160 g sodium acetate ($C_2H_3O_2Na \cdot 3H_2O$) are made up to 1 l with water;
	b) 125 ml conc. acetic acid are made up to 1 l with water; 1 part solution a) and 1 part solution b) are mixed to form the buffer.
Standard zinc solution, β (Zn) = 10 mg/l:	4.399 g zinc sulfate ($ZnSO_4 \cdot 7H_2O$) are made up to 1 l with water. 10 ml of this solution are taken and made up to 1 l with water.

Sodium thiosulfate solution, w ($Na_2S_2O_3$) = 25%

Calibration and Measurement

Aliquots of 0 to 10 ml of the standard zinc solution (corresponding to 0 to 0.1 mg) are taken and treated in the same way as the sample.

The pH of the sample is adjusted to 2 to 3 with hydrochloric acid and 10 ml of this solution are placed in a separating funnel. 5 ml acetate buffer solution and 1 ml sodium thiosulfate solution are added and it is checked that the pH is between 4 and 5.5. 10 ml dithizone solution are added, the mixture shaken for 3 min and the organic phase separated and filtered. The photometric measurement is carried out at 535 nm.

Interfering Factors

Besides zinc, other elements, such as silver, copper, nickel, cadmium and lead, form coloured complexes with dithizone. These elements are almost completely masked by the addition of sodium thiosulfate.

Calculation of Results

The zinc content is obtained by reference to the calibration curve.

6.2 Microbiological Analysis Methods

As water can carry a number of different pathogenic organisms to a large number of consumers, any suspected contamination which could lead to epidemics must be examined.

The monitoring of drinking, industrial, bathing and other waters is carried out using microbiological water tests. In general the tests involve the determination of the total count of virulent organisms and identification of special organisms which are indicative of hygienically suspect contamination (e.g. *Escherichia coli* and coliform bacteria) or are pathogens themselves. Of the pathogens and facultative pathogenic types which can occur in water, the bacteria of the Enterobacteriaceae family are of particular importance. The species *Salmonella, Shigella* and *Escherichia,* the so-called coliform bacteria and *Proteus, Yersinia* and *Erwinia* belong to this family. *Salmonella* and *Shigella* are classed as particularly pathogenic and the others as facultatively pathogenic.

In hygienic water testing mainly the presence of these bacteria is tested. However other hygienically important bacteria can be present, such as *Vibrio cholerae* (causing cholera), *Mycobacterium tuberculosis* (causing tuberculosis), *Clostridium tetani* (causing tetanus) or *Bacillus anthracis* (causing anthrax). In addition the eggs of various parasites can be present in water.

6.2.1 Sample Preparation and Requirements for Microbiological Examination

6.2.1.1 Sampling, Transport and Storage of Water Samples

Sampling is carried out using sterile glass-stoppered bottles where the stopper and neck are covered with aluminium foil to prevent secondary contamination. Before sterilisation sodium thiosulfate solution, c ($Na_2S_2O_3$) = 1 mol/l, should be placed in the bottles to reduce any chlorine present.

0.1 ml of this solution is placed in a 100 ml bottle, 0.25 ml in a 250 ml bottle and 0.5 ml in a 500 ml bottle. Separate bottles should always be used for microbiological tests. They should only be 5/6 filled to facilitate the shaking necessary before testing.

Before sampling from taps, the taps should be fully opened and closed several times to get rid of dirt particles. The tap exit is then flamed until a hissing noise is heard on opening. The water is then allowed to run in a pencil-thick stream for ca. 5 min before the bottle is filled, closed under sterile conditions, and labelled.

In the case of wells with hand pumps the exit is flamed until completely dry. Pumping is then carried out uniformly for ca. 10 min. During this time it should be ensured that the pumped water does not run back into the well nor is allowed to seep into the soil near the well.

In the case of containers or open channels the samples are collected in bottles about 30 cm below the surface with special holders attached to poles of various lengths.

At high external temperatures the samples must be cooled in order to prevent increases in the count after sampling. Samples must be protected against breakage during transport to the laboratory. They should be examined immediately on arrival. If this is not possible they must be stored at ca. 4 °C. Under no circumstances should the period between sampling and testing exceed 48 h (even with cooling). If this is impossible, microbiological tests must be performed on site (e.g. in a mobile laboratory).

6.2.1.2   Technical Requirements in the Laboratory

The Working Area
Micro-organisms are present everywhere so the microbiological working area and the samples must be protected from foreign bacteria (secondary contamination).

Rooms can be disinfected by the action of UV light. The working surfaces should be smooth and easy to clean and disinfect (e.g. stainless steel).

Cleaning and Sterilisation
Sterilisation of apparatus employed in microbiological testing of water is an essential requirement.

Apparatus must be only be sterilised by heating. Plastic culture dishes and other plastic equipment are supplied in prepacked sterile plastic containers.

The following equipment is necessary for sterilisation:

- laboratory oven with or without air circulation,
- steam steriliser,
- autoclave.

New glass apparatus is mechanically cleaned for a short time with mains water and then rinsed with acidified and then distilled water. Used glass equipment which was employed for the examination of impeccable drinking water should be cleaned mechanically with alkali, placed in acidified water, rinsed with distilled water and then dried. All apparatus used in the testing of contaminated water, e.g. glass culture dishes with culture media, should first be sterilised in an autoclave at 120 °C for 20 min and then cleaned as described above. Disposable dishes are boiled or autoclaved for 30 min with water and disinfectant before disposal. Glass equipment which has been cleaned and plugged with cotton wool is sterilised for 2 h at 160 °C in a laboratory oven. Before sterilisation of glass bottles with ground glass joints, a 1 cm wide paper strip is placed between the neck and the stopper. This is removed after sterilisation and the stopper and neck then covered with aluminium foil.

Pipettes are sterilised in pipette tins. The air holes are kept open during sterilisation and then closed. Petri dishes, test tubes and Erlenmeyer flasks are sterilised in wire mesh baskets. With the exception of glass petri dishes all apparatus thus sterilised should remain sterile for a long period. Sterilisation should be repeated after storing for 6 weeks. Plastic culture dishes delivered in sealed sterile plastic bags remain sterile and can be stored for more than 1 year if unopened.

Heat-resistant culture media (e.g. agar agar) are best sterilised using pressurised steam in an autoclave at ca. 120 °C (1 bar gauge) for 20 to 30 min. Thermolabile culture media (e.g. gelatine) are fractionally sterilised, i.e. the media are left in circulating steam for 30 min on 3 consecutive days and incubated in the interim periods at 25 °C. Even heat-resistant bacteria and/or fungal spores are thus destroyed without affecting the growth properties of the medium.

6.2.1.3 Preparation of Nutrient Solutions and Culture Media

General

In the microbiological examination of water culture media based on gelatine or agar agar are used for the determination of the total bacterial count. Liquid nutrient solutions are used as enrichment media and in the 'coloured series' for differentiation between Enterobacteriaceae.

Gelatine and agar agar are the nutrient carriers. They fix the culture medium and thus encourage the isolated growth of bacterial colonies on culture plates. Gelatine is a high molecular weight protein. Gelatine culture media are liquid above 25 °C so they can only be incubated below this temperature. Incubation at 20 to 22 °C is usual. Agar agar is a polysaccharide sulfate ester and is extracted from marine red algae. Agar culture media are liquid at temperatures of around 100 °C and resolidify when the temperature sinks to below 45 °C. Once solidified, agar culture media can still be incubated at temperatures above 45 °C.

Gelatine culture media are liquefied by proteolytic enzymes of some bacteria and moulds. As proteolytic bacteria (predominantly *Pseudomonas* types) mainly live in surface water, the sudden appearance of gelatine liquefiers in subterranean waters is an indication of the influx of surface water.

For the growth of bacteria in culture media optimisation of the pH is particularly important. Dilute hydrochloric acid is used to lower the pH and dilute soda solution or sodium hydroxide is used to raise it.

Preparation Procedures
Gelatine Culture Medium

Meat extract	10 g
Peptone	10 g
Sodium chloride	5 g
Gelatine	120 to 150 g (more in warmer periods)
Demineralised water	1000 ml

The given quantities of the culture medium components are covered with 1000 ml demineralised water in a 2000 ml Erlenmeyer flask. The gelatine is allowed to swell for 1 h at ca. 25 °C and is then dissolved on a water bath at ca. 50 °C. Strong heating before adjustment of the pH should be avoided because gelatine generally reacts as an acid, so at high temperatures it coagulates protein. After dissolving the gelatine the pH of the culture medium is adjusted to 7.2 and two egg whites beaten to a foam are added to clarify it.

The medium is mixed well and heated to 100 °C for 30 to 45 min in a steam steriliser. The coagulated egg white settles at the bottom. The clear culture medium is poured on to a moistened fluted filter paper and is filtered in a hot water funnel or in the steam steriliser. The first part of the filtrate is returned to the filter until the medium runs through clear. A gelatine culture medium should be completely clear and should have a yellowish colour. 10 ml portions of the filtered solution are placed in test tubes, and the latter fitted with cotton wool balls, cellulose stoppers or metal caps and fractionally sterilised in the steam steriliser (3 · 20 min at 24 h intervals). Heating for too long should definitely be avoided as otherwise the gelatine loses its capacity to solidify.

Nutrient Agar

Meat extract	10 g
Peptone	10 g
Sodium chloride	5 g
Agar agar	30 g
Demineralised water	1000 ml

The given quantities of the culture medium components are covered with 1000 ml demineralised water in a 2000 ml Erlenmeyer flask. The mixture is allowed to swell at ca. 25 °C and is then dissolved by heating in a steam steriliser. The pH is adjusted to 7.2 to 7.5 by careful addition of soda solution or sodium hydroxide to the hot liquid culture medium. If turbidity is present the culture medium is clarified with egg white in the same way as gelatine. 10 ml portions of the liquid culture medium are placed in test tubes and the test tubes fitted with cotton wool balls, cellulose stoppers or metal caps. The medium is sterilised either by heating at 120 °C for 15 min in an autoclave or by fractional sterilisation in the steam steriliser (3 × 30 min at 24 h intervals).

The agar culture medium melts at 100 °C and solidifies at 45 °C. It can therefore also be used to cultivate thermotolerant or thermophilic micro-organisms.

Lactose-Peptone Solution

Peptone	20 g
Sodium chloride	10 g
Lactose	20 g
Demineralised water	1000 ml
Bromocresol purple solution:	1 g bromocresol purple in 100 ml water

The given quantities of peptone and sodium chloride are dissolved in 1000 ml demineralised water by heating in a steam steriliser. After leaving for 1 h in the steam steriliser the given quantity of lactose is added and the mixture heated for a further 20 min. The pH is adjusted to 7 by addition of soda or sodium hydroxide solution and 2 ml bromocresol purple are added. This doubly concentrated solution is poured into culture dishes in 10 ml or 100 ml portions for determination of the coli titre by the liquid enrichment method and is sterilised in the autoclave for 20 min at 120 °C.

The normal concentration of lactose-peptone solution is prepared by diluting the nutrient solution prepared as described above with the same volume of demineralised water prior to the addition of the bromocresol purple. The pH is then adjusted and the indicator added. 10 ml portions are placed in test tubes together with a Durham tube and are then sterilised for 30 min in an autoclave at 120 °C.

Durham tubes are similar to 4–5 cm long test tubes with diameters of 6 to 8 mm. Each tube is placed with the open end downwards in the test tube. During sterilisation the air escapes from the Durham tube, which then fills with liquid. Any gas formed during incubation of the inoculated solution collects in the Durham tube.

Endoagar (Lactose-Fuchsin-Sulfite Agar)

Nutrient agar	1000 ml
Lactose	15 g
Alcoholic fuchsin solution	5 ml (10 g diamond fuchsin dissolved in 90 ml ethanol)
Sodium sulfite solution	ca. 25 ml (10 g $Na_2SO_3 \cdot 7H_2O$ in 90 ml water)

1000 ml nutrient agar are placed in a 2000 ml Erlenmeyer flask and are liquefied by heating in the steam steriliser. The given quantities of lactose and fuchsin solution are added and the liquid mixed well. The culture medium thus acquires an intense red coloration. It is then decolourised by addition of sodium sulfite solution. The addition of sodium sulfite must be carried out very carefully. The solution is added until the hot medium only has a faint pink colour (ca. 25 ml sodium sulfite solution). When cold the medium is almost colourless. This is checked by placing a sample of the hot medium in a test tube and cooling in a stream of water until solid.

The endoagar thus prepared contains 3% agar agar and is suitable for preparing subcultures by streaking. The culture medium is light-sensitive and must be stored in a cold, dark location.

For preparing endoagar cultures for use with membrane filters, a smaller concentration of agar agar is desirable to allow for better diffusion of nutrient and indicator. A nutrient agar containing only 1% agar agar is therefore used and must be prepared beforehand. The endoagar with 1% agar agar is too soft for making subcultures by streaking, as the surface of the solidified culture medium is easily damaged by the platinum loop or needle.

Endoagar culture media are generally not placed in test tubes but are fractionally sterilised for 3×20 min at 24 h intervals in Erlenmeyer flasks before pouring directly into sterile petri dishes. The pouring and solidification must be carried out in a darkened room. If such a room is not available the plates must be covered with opaque material during solidification (e.g. several layers of cellulose). The endoagar is stored in a cold, dark location and is predried at 37 °C in the incubator before use. During drying the lid and base of the petri dish are placed face down in the incubator for ca. 30 min at 37 °C.

Culture Medium for the Selective Detection of Enterococci (according to Slanetz and Bartley)

Tryptose	20 g
Yeast extract	5 g
Glucose	4 g
Disodium hydrogen phosphate	
($Na_2HPO_4 \cdot 2H_2O$)	4g
Sodium azide (NaN_3)	0.4 g
Triphenyltetrazolium chloride (TTC)	0.1 g
Agar agar	10 g
Demineralised water	1000 ml

The above ingredients with the exception of sodium azide and TTC are dissolved in the water by careful heating to boiling point. After cooling to 46 °C the solution is treated with 10 ml 1% aqueous sterile filtered TTC solution and 4 ml 10% aqueous sterile filtered sodium azide solution. No further sterilisation is carried out. The culture medium is poured directly into sterile petri dishes and is allowed to solidify on a horizontal surface. This culture medium is particularly suitable for testing water by the membrane filter method but is only stable for a limited time.

Culture Medium for Selective Detection of Pseudomonas aeruginosa (Cetrimide Agar)

Peptone	20 g
Magnesium chloride ($MgCl_2$)	1.4 g
Potassium sulfate (K_2SO_4)	10 g
N-cetyl-N,N,N-trimethyl-	
ammonium bromide	0.5 g
Agar agar	13.6 g
Glycerol, doubly distilled	10 ml
Demineralised water	1000 ml

The above ingredients with the exception of the glycerol are treated with 1000 ml demineralised water in a glass flask and shaken vigorously. The mixture is allowed to swell for 15 to 30 min at 20 to 25 °C, during which time the nutrient mixture must be kept still. 10 ml doubly distilled glycerol are then added. The mixture is heated to boiling with frequent swirling. The pH is adjusted to 7.3 to 7.4 and 10 ml portions of the selective agar are placed in test tubes and sterilised for 15 min in an autoclave at 120 °C. Alternatively the whole quantity is sterilised in the flask and then poured into sterile petri dishes before solidification.

Ready-made Culture Media

Dried culture media in powder form containing all the components necessary for cultivating the respective micro-organisms are commercially available. A defined quantity of the powder must be weighed out and a defined quantity of demineralised or mains water poured over it according to the manufacturer's recipe. The medium is then generally dissolved by heating and the pH adjusted to the required value. After transferring to test tubes the medium is either sterilised in an autoclave or subjected to fractional sterilisation in the steam steriliser. Adjustment and subsequent checking of the pH is recommended. Some manufacturers also supply ready-made culture media in tablet form. The tablet corresponds to a particular quantity of water so weighing is not necessary.

6.2.2   Tests

6.2.2.1   Total Bacterial Count

The total bacterial count using indirect culture methods cannot cover all organisms, as each culture medium is selective. The principle of the method is that each bacterium capable of multiplying is visible on a nutrient substrate as an individual colony. For mixed populations the validity of this principle is very limited.

The total bacterial count is the number of colonies under a magnification of 6 to 8 which have developed under defined conditions. It provides information on the degree of contamination of water by micro-organisms and, in particular, sudden bacterial invasions.

Apparatus
Basic microbiological equipment

Reagents and Solutions
Standard culture medium: culture medium as described in Section 6.2.1.3 or DEV
                                  nutrient agar (Merck) or Plate Count Agar (Oxoid) is
                                  available in sterile 10 ml portions.

Measurement
1 ml of a well mixed water sample is pipetted into a sterile petri dish and mixed with sterile nutrient gelatine or agar. Where high counts are expected, diluted cultures should be prepared with sterile water (e.g. 1:100,1:1000 etc). The gelatine is liquefied at 35 °C in a water bath and cooled to ca. 30 °C before pouring into the petri dishes.

Tubes of nutrient agar are liquefied in boiling water and cooled to 46 °C before pouring out. The openings of the tubes must first be flamed. The petri dishes are swirled in a figure-of-eight movement to achieve thorough mixing and are then left on a horizontal surface. Nutrient gelatine solidifies below 25 °C and so cooling may be necessary.

The culture dishes with the solidified media are incubated at 20 and/or 37 °C for 44 ± 4 h with the culture medium layer uppermost.

Calculation of Results
The visible colonies are counted using a 6 to 8 x magnifying glass. With agar culture media and an incubation temperature of 37 °C an initial count can be performed after 20 h. A Wolffhügel counting plate or another suitable counting instrument is used for heavily populated plates.

The total bacterial count is based on 1 ml of the water sample. For values > 100 it is rounded off to the nearest 10 and for values > 1000 to the nearest 100. The culture medium used, incubation time and temperature must also be given. Example: total bacterial count (gelatine, 44 h, 20 °C): 90 colonies/ml.

Other (but not official) methods are:

Total Count Determination by Inoculation of Agar Plates:
Sterile agar plates with a dry surface are inoculated with a known sample volume evenly distributed over the surface with a spatula (e.g. a Drigalski spatula).

Miles and Misra Dropping Method:
Five drops of sample are allowed to fall from a pipette at a height of ca. 20 mm on to a well dried agar plate. After incubation the total count per ml is calculated from the mean count of colonies per plate.

Membrane Filter Method:
Direct cultivation on 0.45 µm membrane filters is possible. If such a filter is placed on a plate with a suitable solid culture medium (nutrient agar or cardboard filter discs impregnated with nutrient solution), the nutrients diffuse through the filter allowing colonies to be formed on the filter surface.

The advantage of this method is that even at very low micro-organism concentrations the bacterial count can be determined by increasing the sample volume. Growth inhibitors dissolved in the sample are removed in the filtration and therefore cannot affect the subsequent growth of micro-organisms on the culture medium.

Dip Slide Method:
A sterile glass plate with the dimensions of a normal microscope slide (26 x 76 mm) is fixed to the underside of the lid of a suitable sterile cylindrical container (Fig. 36). The slide is coated with the culture medium on one or both sides. Such slides are commercially available.

The inoculation is carried out by dipping the slide briefly into the sample or diluted sample. Excess sample is drained off at the bottom of the slide by blotting on filter paper.

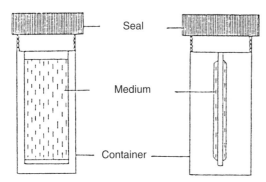

Seal

Medium

Container

Fig. 36: Apparatus for determination of total bacterial count by the dip slide method

Approximately the same sample volume always remains on the surface of the coated slide. For incubation the slide is placed back in the cylindrical container. After incubation the bacterial count is determined by comparing the colony density with standard slides. The colonies may also be counted, but there is no advantage in this because of the low precision of the method. The dip slide method is suitable for colony counts of ca. 10^3/ml or more.

6.2.2.2   Escherichia coli and Coliform Bacteria

Liquid Enrichment Procedure:

If it is only necessary to test whether *Escherichia coli* and/or coliform bacteria are present in 100 ml water, the following procedure is adequate: 100 ml sample are mixed with 100 ml of a doubly concentrated lactose-peptone solution and incubated at 37 °C for 20 ± 4 h. The mixture is then tested for production of acid and gas. If acid or gas are not produced, the water satisfies drinking water requirements regarding *E. coli* and coliform bacteria and the test procedure can be stopped. If there is lactose fermentation with acid and gas production, it must be determined whether the organisms responsible are *E. coli,* coliform bacteria or bacteria of a different group.

For this purpose, a small quantity of the turbid fermented lactose-peptone solution is taken on a sterile platinum loop and fractionally streaked on endo agar. Fractional streaking means that the loop is not streaked over the whole endo plate, but only a single layer of organisms is applied to the surface of the medium near the petri dish edge. A second sterile loop is then used to streak part of this material at right angles and over a third of the medium surface. After turning the plate through a further 90°, a newly sterilised loop is used to transfer further material to the part of the surface not yet streaked. In this way single colonies can be obtained and identified in the so-called 'coloured series'.

Moist dark red colonies with a golden shimmering metallic tinge are suspected of being *E. coli.* Coliform bacteria grow as moist red colonies with constant or varying metallic tinges and with or without slime formation.

To apply a coloured series commercially available kits such as API, Enterotube or Titertek are used. These are preprepared culture media, which are inoculated with material from a single colony and are then incubated. Handling and incubation are carried out according to the manufacturer's instructions. Evaluation is often carried out by obtaining a numerical code based on the occurrence of positive or negative metabolic reactions with the individual culture media. Reference to a code number in a list supplied allows the type of organism to be identified.

Where such kits are not available, the identification media must be prepared, inoculated and the organism identified according to the results obtained.

Table 26: Determination of biochemical properties on different media of a 'coloured series'

Culture medium	Incubation temperature (°C)	Positive reaction	Negative reaction
Nutrient agar plate	37	Single typical colonies only	Morphologically distinguishable colonies
Simmons citrate agar angled	37	Growth with colour change from green to blue	No growth, no colour change
Koser citrate solution	37	Turbidity caused by bacterial growth	No growth, clear, no turbidity

Continuation **Table 26:**

Culture medium	Incubation temperature (°C)	Positive reaction	Negative reaction
Glucose-peptone solution culture a) culture b)	37 44	Turbidity, gas development, colour indicator changes from purple to yellow	No growth, no gas development, no colour change
Lactose-peptone solution	44	Turbidity, gas development, colour change from purple to yellow	No growth, no gas development, no colour change
Neutral red mannitol broth	44	Turbidity, gas development, colour change from red to yellow	No growth, no gas development, no colour change
Urea Kligler agar	37	Slant surface: colour change from red to yellow, as a result of acid production; inoculation: gas development, blackening by H_2S, NH_4 formation by urea degradation	No gas development, no colour change through acid production, no blackening by H_2S, no NH_4 formation
Tryptophan-trypton broth	37	Growth with turbidity, red coloration on addition of indole reagent	No growth, no red coloration with indole reagent
Buffered nutrient solution	37	Solution is divided between two sterile test tubes a) methyl red sample: colour change from yellow to red b) Voges-Proskauer reaction (addition of KOH + creatine): red coloration after 1 to 2 min	Indicator remains yellow No colour after 2 min
Nutrient gelatine	20–22	Liquefaction in inoculation area	No liquefaction

Determination of Biochemical Properties:

The culture media listed in Table 26 are inoculated as a coloured series and incubated at the given temperatures for 20 ± 4 h.

For a pure culture the nutrient agar plate must only show typical colonies with the same appearance. If different colonies are visible, then the starting colony was of a mixed variety and therefore unsuitable for differentiation purposes. In such cases new subcultures must be streaked on endo agar. If a pure culture is present on the nutrient agar plate, the cytochrome oxidase reaction is carried out. 2 to 3 drops of Nadi reagent (N-tetramethyl-p-phenylenediamine dihydrochloride) are placed on the colonies using a dropper. A positive reaction is shown by the colonies becoming violet blue in 1 to 2 min. No colour change is visible in the negative case.

A positive cytochrome oxidase reaction shows that *E. coli* and coliform bacteria are not present. With a negative cytochrome oxidase reaction the coloured series needs to be interpreted according to the scheme shown in Table 27.

Table 27: Biochemical properties of *Escherichia coli* and coliform bacteria

Reaction	*E. coli*	*Enterobacter*	*Klebsiella*	*Citrobacter*
Glucose fermentation				
(37 °C)	+	+	+	+
(44 °C)	+	+/–	+/–	+/–
Lactose fermentation (44 °C)	+	+/–	+/–	+/–
Mannitol fermentation (44 °C)	+	+/–	+/–	+/–
Citrate degradation	–	+	+	+
Indole formation	+	–	–	+/–
Methyl red test	+	–	–	+
Voges-Proskauer reaction	–	+	+	–
Urea degradation	–	+/–	+	+/–
H$_2$S formation	–	–	–	+/–
Gelatine liquefaction	–	–	–	–

+ positive
– negative
+/– differing behaviour of various species

For the determination the following is important:

E. coli grows at 44 °C and ferments glucose, lactose and mannitol with gas formation; it forms indole and has a positive methyl red test, whereas the Voges-Proskauer reaction, urea degradation and hydrogen sulfide production are negative. Citrate is not degraded.

Coliform bacteria generally do not grow at 44 °C but at 37 °C. They can degrade citrate and differ from each other through H$_2$S formation, urea degradation, indole formation, methyl red reaction and the Voges-Proskauer reaction.

Membrane Filter Method:
A large quantity of water (100 ml or more) is filtered through a sterile 0.45 µm membrane filter, which is then incubated in lactose-peptone solution at 37 °C or placed free of air bubbles on endo agar or endo nutrient cardboard discs. After incubation any suspected colonies are identified as described above. Turbid samples should not be membrane filtered as the pores become blocked and subsequent inhibition of growth on the filter surface may occur.

Coli Titre:
The coli titre gives the smallest volume of sample in which one bacterium of the Enterobacteriacea family is just detectable. Different volumes of water (e.g. 100 ml, 10 ml, 1 ml, 0.1 ml etc.) are incubated in lactose-peptone solutions of corresponding concentrations. The smaller the sample volume is, in which a reproducible bacterium can be detected, the higher the bacteria concentration of the sample is (Table 28).

Table 28: Relationship between coli titre and bacterial density

Titre (ml)	Bacteria/ml
1000	0
100	0
10	0
1.0	1–9
0.1	10–99
0.01	100–999
0.001	1000–9999

Examples:
'In 100 ml water *E. coli* and coliform bacteria not detectable'.
'*E. coli* is detectable in 0.1 ml water'.

6.2.2.3   Other Hygienically Important Micro-organisms

Faecal Streptococci (Enterococci):
Besides *E. coli* and coliform bacteria, these are indicators of faecal contamination of water as they are normally present in the intestinal tract of humans and animals. They rarely multiply in water but possess an above average resistance to heat, alkali and salts. They thus grow at pH 9.6 and temperatures of 10 to 45 °C in a medium with 6.5% sodium chloride content and are not inhibited by azide.

The microbiological determination can be performed either with the liquid enrichment method in azide-dextrose broth with subsequent application on blood-azide-agar or by membrane filtration. A more accurate identification can be carried out microscopically and by serological methods (References: Suess, 1982).

Pseudomonas aeruginosa:

This organism is also a faecal indicator and therefore a facultative human pathogen, often causing wound, eye or ear infections. It is extremely resistant to antibiotics and is often found in hospitals. It has been isolated from water which has passed tests for *E. coli* and coliform bacteria.

Microbiological tests are performed by liquid enrichment in malachite green broth and subsequent inoculation on a cetrimide medium or using membrane filtration, the filter being placed on cetrimide agar. After incubation at 37 °C these organisms form the green pigments fluorescein and pyocyanine. The cultures also have a characteristic sweetish aromatic odour.

Clostridium perfringens:

This anaerobic spore-forming bacterium is present in the faeces of warm-blooded animals, but in much lower concentrations than *E. coli*. As clostridial spores can survive for a long time in the environment, their detection in the absence of *E. coli* or coliform bacteria indicates an older water contamination and not a current one.

Detection is carried out by liquid enrichment: 20 to 100 ml sample are anaerobically incubated with an equal volume of doubly concentrated dextrose – iron citrate – sodium sulfite broth. Blackening of the liquid medium indicates a positive reaction. In the membrane filter method the water is drawn through the filter, the reverse side placed on dextrose – iron citrate – sodium sulfite agar and incubated anaerobically at 37 °C.

Parasites, in particular Worm Eggs:

Untreated waste water is the main source of danger for the transfer of worm parasites. Worms, their larvae and other developmental stages can get into water with the faeces of humans and domestic animals. The efficiency of eradication of worm eggs in a waste water treatment plant must therefore be determined periodically before and after processing. Worm eggs in sewage sludge can survive the treatment process so its use as a fertiliser may present a risk of infection. Such a risk also exists in the application of untreated waste water or dung to fruit and vegetables and in bathing in polluted water and untreated swimming baths.

Water treatment plants fed by surface water require regular examination for parasite eggs, especially when cattle are reared near the water.

Enrichment of worm eggs in saturated sodium chloride solution:

A sludge or water sample of ca. 1 g is mixed in a ratio of 1:27 with saturated sodium chloride solution (377 g NaCl in 1 l water). The solution must be added slowly, initially dropwise with stirring so that a fine, even bilayer results which is left to stand for 15 to 20 min. The eggs of all nematodes and certain cestodes float in saturated sodium chloride solution, unlike other sludge components, because of their lower specific gravity. Small beakers are employed to concentrate the rising worm eggs over as small an area as possible. A wire loop (1 cm diameter) bent at right angles to the stem is placed flat along the surface and is used to remove worm eggs without contaminating particles. The drops are transferred to a slide and examined microscopically without a cover glass. This method is unsuitable for the detection of trematode eggs. Using a saturated zinc chloride solution with a higher specific gravity the recovery can be significantly increased, even for trematode eggs.

6.2.2.4   Growth Inhibition of Luminescent Bacteria

Marine luminescent bacteria of the family *Vibrionaceae* are related to the terrestrial enterobacteria, among them *E. coli*. The decrease in bioluminescence during the standardised test time after comparison with unaffected luminescent bacteria samples is used as a measure for the toxic inhibition of a water sample. Luminescent bacteria of the type *Vibrio fischeri* (also known as *Photobacterium phosphoreum*) are used for the test. Preserved bacteria are commercially available. A luminometer is used as the measuring instrument and measures the intensity of the light emitted by the luminescent bacteria. An important advantage of the method over other biological tests is the short test duration of 30 min.

The procedure is based on that described in DIN 38 412, part 34 and the extension DIN 38 412, part 341. The luminescent bacteria stored at −18 °C are reactivated with the dilution solution and can be used after ca. 30 min. The luminous intensity of the control mixture is determined before addition of the sample. The sample (in several degrees of dilution) and the bacteria are incubated and the luminous intensities of all the mixtures are determined.

Apparatus
Freezer
Thermostat equipment in the range 15 ± 0.2 °C
Luminometer with measuring cell for 15 ± 0.2 °C, with glass cuvettes
pH meter with electrode
Micropipettes

Chemicals
Preserved luminescent bacteria: stored at −18 to −20 °C
Sodium chloride solution: 20 g NaCl are dissolved in 1 l water.
3,5-Dichlorophenol
Zinc sulfate ($ZnSO_4 \cdot 7H_2O$)
Potassium dichromate

Measurement
The pH of the sample is adjusted to 7 and sodium chloride added until a concentration of ca. 20 g/l is reached. A series of dilute solutions is prepared from this solution (dilution ratios: 1:2, 1:4 etc.). 1 ml water at 3 °C is then poured rapidly over a portion of preserved bacteria and the solution kept at 3 °C as a stock solution.

After 5 min a test suspension is prepared by adding 10 µl stock solution to 500 µl sodium chloride solution at 15 °C and shaking.

The luminous intensity I_0 of a freshly prepared control mixture consisting of 0.5 ml test suspension and 0.5 ml sodium chloride solution is then measured immediately in the luminometer. Various test mixtures consisting of 0.5 ml test suspension and 0.5 ml of each of the different diluted samples are prepared. After leaving to stand for 30 min, the luminous intensities I_t of all sample mixtures and the control mixture are measured.

Calculation of Results

The chemicals 3,5-dichlorophenol (6 mg/l), zinc sulfate (0.8 mg/l) and potassium dichromate (14 mg/l), which should cause 20–80% inhibition at the concentrations given, are used as reference substances for determining the activity of the luminescent bacteria.

The correction factor for the luminous intensity (given in relative luminosity units) is calculated according to:

$$f_{k30} = \frac{I_{k30}}{I_0}$$

I_{k30} luminous intensity of the control mixture after 30 min

I_0 luminous intensity of the test suspension before addition of the diluted sample

From this the theoretical luminous intensity of a test mixture containing no inhibiting substances can be calculated:

$$I_{c30} = I_0 \cdot f_{k30}$$

f_{k30} is the mean value of several f_{k30} determinations.

The inhibiting activity H_{30} of a test mixture is given by:

$$H_{30} = \frac{I_{c30} \cdot I_{T30}}{I_{c30}} \cdot 100(\%)$$

I_{T30} intensity of luminescent bacteria in the test mixture after 30 min.

6.2.2.5   Respiration Inhibition

Respirometric procedures, which are suitable for determining biodegradability, can also be used to measure toxicity towards micro-organisms. Toxic substances can, for example, inhibit the degradation of domestic or industrial waste water in treatment plants, if certain concentrations are exceeded.

For degradation the same quantity of oxygen is required within a defined period of time as is consumed in gas metabolism by biochemical conversions in the sample. This oxygen demand is known as biochemical oxygen demand (BOD). BOD depletion curves are mostly discontinuous. They show that readily degradable organic materials are first attacked followed by the less readily degradable ones. Substrate respiration is often indicated by a plateau (Fig. 37).

A simple procedure for the determination of respiration inhibition in the degradation of peptone, the Zahn/Wellens modification of the OECD 301 E open aeration test, and the determination of the short term respiration inhibition of activated sludge according to OECD 209 are described below.

a) Respirometric Inhibition Test:

This test can be used to determine the inhibiting effect of substances on biochemical degradation. The following three mixtures are prepared to determine the toxic or inhibiting effect of components of waste water on micro-organisms:

– only sample solution,
– only readily degradable substance,
– sample solution and readily degradable substance together.

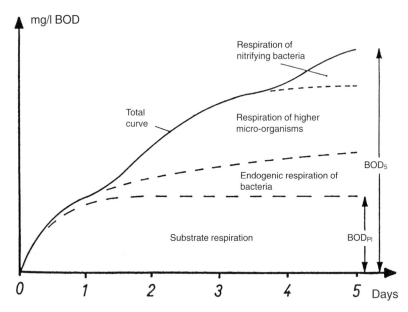

Fig. 37: BOD depletion as a function of time

Respirometric depletion takes place in closed vessels. If the oxygen concentration in the solution is lowered by biochemical degradation, gaseous oxygen passes into the aqueous phase and the partial pressure of oxygen decreases. This pressure drop is measured with simple instruments manually by reading a U-tube manometer filled with mercury or a digital pressure sensor connected to the container. More advanced instruments are also on the market, among them those with oxygen delivery (e.g. SAPROMAT®).

The carbon dioxide produced in the oxidation of organic substances is removed from the gas chamber by absorption in potassium hydroxide. A mineral nutrient solution is used for dilution. The inoculation solution is a coarsely filtered run-off from a municipal sewage plant.

Apparatus
Manometric BOD measuring instrument with eight to twelve 500 ml vessels and a magnetic stirrer
Alternatively: automatic BOD measuring instrument (e.g. SAPROMAT®)

Reagents and Solutions

Dilution water:	in each case 1 ml of solutions a) to d) is made up to 1 l with water:
	a) 8.5 g KH_2PO_4, 21.8 g K_2HPO_4, 33.4 g $Na_2HPO_4 \cdot 2H_2O$, 2.5 g NH_4Cl are dissolved in 1 l water; pH = 7.2.
	b) 22.5 g $MgSO_4 \cdot 7H_2O$ in 1 l water
	c) 27.5 g $CaCl_2 \cdot 6H_2O$ in 1 l water
	d) 0.25 g $FeCl_3 \cdot 6H_2O$ in 1 l water.
Inoculation solution:	run-off from a sewage plant, filtered through fluted filter paper
Peptone solution:	1 g peptone is dissolved in 1 l water.

Potassium hydroxide solution, w (KOH) = 45%

Measurement

Defined quantities of substrate (1. sample solution in various concentrations, 2. peptone solution, 3. sample solution in concentrations as in 1 + peptone solution) are placed in 500 ml graduated vessels. For the measuring range see the manufacturer's instructions. After adjusting all solutions to pH 7, 10 drops of inoculation solution are added to each one. The absorption vessel is filled with potassium hydroxide, the measuring cell closed, and the zero point adjusted after ca. 30 min.

With a simple instrument the manometer is read four to five times daily. The length of the test is usually 5 days (BOD_5) or 28 days if the OECD 301 F degradation test is formally carried out (BOD_{28}).

Calculation of Results

At the end of the test three BOD curves are obtained for a given sample concentration: a) that of the sample (in different concentrations where necessary), b) that of the peptone solution and c) that of the sample/peptone mixture. The area under the curve is proportional to the oxygen consumption. The area for the mixture with the sample is compared with that for the mixture without the sample to determine the inhibition of respiration.

Figure 38 shows an example for a waste water sample. The addition of the two curves for the waste water and peptone gives the upper (theoretical) curve. The curve for the mixture of waste water and peptone clearly lies below it. The area between these last two curves can be interpreted as the inhibition area and expressed as a percentage of the total area. Integration is performed simply by cutting out and weighing the individual areas of the recorder paper or, as is preferred with a series of measurements, using an integrator or integration calculation program.

b) Open Aeration Test:

In this procedure the degradation of organic substances and substance mixtures and its inhibition are followed by continuing measurement of the total parameter 'dissolved organic carbon' (DOC). According to OECD 301 E the formal test duration is 28 days, but it can be shortened depending on the requirements of the inhibition test.

Apparatus
DOC/TOC measuring instrument
2 l wide-necked bottles with conical shoulders
Aquarium pump with aeration stones
Membrane filtration apparatus with 0.45 µm filter

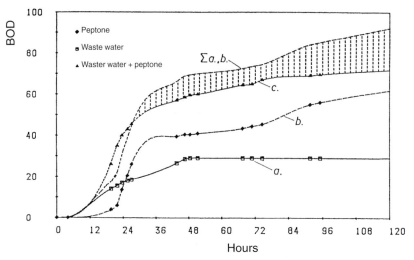

Fig. 38: Example of inhibition of BOD in a respiration inhibition test

Reagents and Solutions

Dilution water:	4 ml of solution a) and 1 ml of solution b) are made up to 1 l with water;
	a) 109.5 g $CaCl_2 \cdot 6H_2O$ in 1 l water;
	b) 123.3 g $MgSO_4 \cdot 7H_2O$ in 1 l water.
Nutrient solution:	38.5 g NH_4Cl and 9.0 g $Na_2HPO_4 \cdot 7H_2O$ are dissolved in 1 l water.
Peptone solution:	1g peptone is dissolved in 1 l water.
Inoculation solution:	run-off from a sewage treatment plant, filtered through fluted filter paper.

Measurement
Defined quantities of substrate (1. sample solution in various concentrations, 2. peptone solution, 3. sample solution in concentrations as in 1 + peptone solution) are placed in 2 l wide-necked bottles with conical shoulders. After addition of 5 ml nutrient solution and 1 ml inoculation solution to each one, the volume is made up to 2 l with dilution water and the pH adjusted to 7.0. The solutions are then aerated with an air throughput of ca. 15 l/h.

A sample of ca. 10 ml is taken after 6 h to begin with, then each day, and after 10 days ca. every 3 days. It is filtered through a 0.45 µm membrane filter and the DOC concentration measured.

If all the samples are to be tested at the same time, they should be frozen at –20 °C in poly-ethylene containers.

Calculation of Results

The DOC content of the sample is calculated according to:

$$DOC_t = (DOC_{M,t} - DOC_{0,t}) \cdot \frac{2000}{2000 - V_S}$$

DOC_t DOC of substrate and sample at time t, mg/l
$DOC_{M,t}$ measured value at time t, mg/l
$DOC_{0,t}$ blank value (dilution water + inoculation solution) at time t, mg/l
V_S sum of all the sample volumes taken for the measurement, l

If only the biodegradability of a sample is to be determined, and not its toxicity, only the sample and the blank solution are used for the test.

c) Respiration Inhibition of Activated Sludge

OECD 209 describes the testing of the inhibiting action of waste water on activated sludge respiration. The oxygen depletion of the micro-organisms contained in the activated sludge is determined in a defined period of 3 h, once after the addition of a defined synthetic waste water sample and then after addition of the actual waste water sample. The inhibiting action of the waste water is given as the EC_{50} value (= dilution at which an inhibition of 50% compared with the control occurs), when different degrees of dilution are being tested. An inhibiting substrate, e.g. a solution of 3,5-dichlorophenol, is used as a control substance.

Apparatus
1 l beakers
500 ml wide-necked bottles with conical shoulders
pH meter with electrode
Oxygen measuring instrument with electrode
Aquarium pump with aeration stones

Reagents and Solutions

Synthetic waste water:	the following ingredients are dissolved in 1 l water: 16 g peptone, 11 g meat extract, 3 g urea, 0.7 g NaCl, 0.4 g CaCl$_2$. · 6H$_2$O, 0.2 g MgSO$_4$ · 7H$_2$O, 2.8 g K$_2$HPO$_4$. The solution should be stored in the dark at 4 °C for not longer than 1 week.
Control solution:	0.5 g 3,5-dichlorophenol are dissolved in 10 ml sodium hydroxide, c (NaOH) = 1 mol/l; the solution, diluted with 30 ml water and sulfuric acid, c (H$_2$SO$_4$) = 0.5 mol/l, is added

| Activated sludge: | with stirring until precipitation is visible (ca. 8 ml acid). The solution is then made up to 1 l. The pH should be 7–8. ca. 2 l activated sludge are taken from a municipal sewage treatment plant with the smallest possible proportion of industrial waste water. Ca. 10 ml of the homogenised sludge is taken and the dry substance is determined. Depending on the result, the sludge is diluted to 2 to 4 g dry substance per l (corresponding to 0.8 to 1.6 g/l in the subsequent mixture). If the sludge is not used on the day it was taken, 50 ml synthetic waste water is added and the mixture aerated at 20 °C. If necessary the pH is buffered to 6 to 8 using sodium hydrogen carbonate. |

Measurement

For the control mixture C_1 16 ml synthetic waste water are diluted to 300 ml with water. 200 ml activated sludge are added and the mixture is aerated in a beaker for 3 h with an air throughput of 0.5 to 1 l/min at 20 ± 2 °C.

After 15 min the second mixture is prepared. 250 ml of the waste water sample are added to 16 ml synthetic waste water and the solution made up to 300 ml. 200 ml activated sludge are added and the mixture aerated as before. This procedure is repeated at 15 min intervals with smaller volumes of the waste water sample (e.g. 125 ml, 62.5 ml made up to 300 ml with water) and then with different volumes of the control solution (measuring range for 3,5-dichlorophenol: 5 to 30 mg/l in the final mixture). Finally a control mixture C_2 is analysed in the same way as C_1.

3 h after the first analysis solution C_1 is placed in a 500 ml wide-necked bottle with a conical shoulder. An oxygen electrode is introduced immediately and is sealed with a stopper so that no air bubbles remain. The decrease in oxygen concentration in the flask is measured over 10 min either using a recorder-printer or is read off every min and recorded. These depletion measurements are repeated every 15 min with all the mixtures so that they are all measured after 3 h aeration time.

Calculation of Results

The oxygen depletion of each mixture is taken from the print-out or the measurement records. The time interval for the decrease in the oxygen concentration from 6.5 mg/l to 2.5 mg/l (if possible) is determined and converted into mg/l · h (at lower depletion the whole 10 min measuring time is determined). The measurements of C_1 and C_2 should not differ from one another by more than 15%.

The inhibiting action of the water sample at a given concentration is calculated as follows:

$$\text{Inhibition } (\%) = \left(1 - \frac{2P}{C_1 + C_2} \right) \cdot 100$$

P oxygen depletion of the water sample at a given dilution
C_1, C_2 oxygen depletion in controls 1 and 2

The percentage inhibition is calculated for each sample dilution tested, plotted on logarithmic paper against the dilution and the EC_{50} determined by interpolation.

6.3   Methods for Soil Analysis

Within the framework of this book it was obvious that selected methods for soil examination should be included, because problems regarding soil fertility and melioration are often associated with the quality and economy of the water. Detailed explanations based on soil science are not given, emphasis being placed here on practical elements of the analysis. Some methods of soil analysis can also be found under the description of the individual parameters in Chapter 5 and Section 6.1.

6.3.1   Sample Preparation

After collection soil samples are dried as rapidly as possible, provided that particular measurements which depend on the water content are not required. Drying must be carried out carefully to avoid secondary reaction processes. Large lumps of soil are crushed and roots and animals removed. Fine soil is used for most soil analyses. The unprocessed soil is sieved through a 2 mm sieve. Soils rich in clay are sieved before they are completely dried. For some analyses (e.g. testing for metals) the sample must be ground until it is as fine as dust.

The most important stages of sample preparation are described below.

6.3.1.1   Sorting

Before laboratory tests heterogeneous soil samples must frequently be sorted, reduced in size or sieved in order to obtain a homogeneous sample. The individual components thus obtained can then be weighed separately, further processed, analysed and the result calculated for the whole population or the sorted parts of it. Inert coarse particles > 2 mm (stones, pebbles, rubble) are usually removed. Their proportion by weight in the sample should be noted, however, and if necessary taken into account when presenting the results. For particle size determination see Section 6.3.2.2.

6.3.1.2   Drying

In practice a differentiation is made between

– air drying,
– drying at elevated temperatures,
– freeze drying.

Air drying is most commonly used and is prescribed in the Sewage Sludge Ordinance. Drying in a circulating air drying oven at a maximum of 40 °C accelerates the drying process, whereby the samples are spread out as thinly as possible and turned over several times. Larger lumps are crushed. A high clay content usually leads to caking so these samples need to be ground up and redried. The standard procedure at elevated temperature involves drying at 105 °C. If the moisture content of the soil is to be determined as completely as possible, it should first be dried for several hours at 105 °C and then at 180 °C. Strongly bound residual water and the water of crystallisation of most salts is thus liberated.

Freeze drying is used less often, e.g. in the examination of biochemical soil parameters.

6.3.1.3   Sample Reduction

After sorting it is necessary to reduce the size of the soil sample to a partial sample which can be analysed. For some chemical soil analyses extensive sample reduction is necessary.

Soil samples are mixed by repeated shovelling using the cone method (Fig. 39). The coarse particles run down the side of the cone to the base. The cone can be divided into four parts crosswise. The two diagonally opposite parts are combined to form a new cone and treated in the same way. The remainder is discarded. After the sample reduction step the two quarters to be used for further processing are ground.

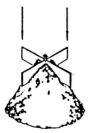

Fig. 39: Cone formation and division into quarters

Reducing the size of samples between the individual mixing steps is not usually carried out in a single step. The sample reduction technique chosen depends on the size, structure and hardness of the material. Soils with a particle size of ca.10 mm before sample reduction can be first ground to an intermediate size of 4 mm using a hammer mill or a rotary ball mill. During the subsequent fine grinding step attention must be paid to the increased risk of contamination by metal particles rubbed off through abrasion in the mills, particularly at particle sizes < 0.5 mm. If necessary mills made of the metals being investigated should not be used. Particle sizes of < 0.25 mm can be reached using planetary ball mills with tungsten carbide inserts or mortar mills with porcelain or agate inserts. If particle sizes of well below 0.25 mm are to be achieved, a fast running centrifugal mill (10 000 to 20 000 min^{-1}) is usually used. Exchangeable ring sieves define the final particle size. Heavy metal-free inserts are also available for these types of mill.

6.3.1.4   Digestion

The type of digestion chosen depends on the nature and composition of the soil sample, the properties of the elements to be determined and the analysis procedure.

Simple and efficient procedures are always preferred to the more expensive ones if no or very few methodological errors are expected. For the determination of heavy metals in many soils and sewage sludges a total digestion of the mineral matrix is not necessary, as the heavy metals are predominantly bound to the surfaces of the particles. In other soils, however, most of the heavier metals are strongly bound within the mineral matrix. In this case a total digestion is necessary.

If only alkali and alkaline earth metals are to be determined, simple ashing is sufficient, possibly with the addition of oxidising agents. Some tried and tested digestion procedures for the group determination of some important main and trace metals are described below.

Digestion for the Total Determination of Ca, Mg, K, Na, Fe, P

0.5 g air-dried ground fine soil are heated to red heat in a platinum crucible, cooled, moistened with water and then treated with 1 ml perchloric acid, w ($HClO_4$) = 60%, and 10 ml hydrofluoric acid, w (HF) = 40%. The contents of the crucible are allowed to fume at ca. 180 °C on a sand bath. After cooling 15 ml hydrochloric acid, w (HCl) = 10%, are added and the closed crucible is heated to dryness. If the solution is not clear, the fuming process with $HF/HClO_4$ must be repeated. The solution is then transferred to a 100 ml volumetric flask.

Digestion with Aqua Regia

In Germany the Sewage Sludge Ordinance of 1992 prescribes the digestion of sewage sludge and soils with aqua regia according to DIN 38414, part 7 for the determination of metals.

3 g dried sample are placed in a cylindrical 200 ml glass vessel fitted with an air or water condenser and 21 ml hydrochloric acid, w (HCl) = 35%, and 7 ml nitric acid, w (HNO_3) = 65%, are added. The mixture is kept at room temperature for several hours or overnight. It is then heated to boiling point for 2 h on a metal heating plate. The contents of the vessel are made up to 100 ml with water added down the condenser. After sedimentation or filtration of the solution the elements Cd, Cr, Cu, Hg, Ni, Pb, Zn, K, Na, Ca, Mg and P can be determined.

Total Digestion in a Pressurised Vessel

For soils and sewage sludges hydrofluoric acid must be used together with oxidising acids to digest silicate-containing material. A suitable procedure is as follows:

300 mg sample are treated with 2 ml nitric acid, w (HNO_3) = 70%, 6 ml hydrochloric acid, w (HCl) = 35%, and 3 ml hydrofluoric acid, w (HF) = 40%. The mixture is heated in a PTFE pressurised digestion vessel in a microwave oven programmed for 2 min at 144 W, 3 min at 280 W, 3 min at 420 W and 1 min at 560 W. Between each heating step there is a cooling period of 1 min. After cooling, 15 ml saturated boric acid is added to bind excess hydrofluoric acid and the volume is then made up to 100 ml with water. Loss of mercury during the digestion is prevented using this method.

Digestion for Weatherable Ca, K and P Compounds

10 g air-dried fine soil are calcined in a large porcelain crucible for 1 h at 500 °C. After cooling 50 ml hydrochloric acid, w (HCl) = 30%, are added and the mixture carefully heated on a sand bath with the crucible covered with a watch glass. The mixture is then filtered into a 100 ml volumetric flask, the solid material washed with water, and the solution in the flask made up to the mark. To remove chloride, which can interfere with phosphorus determination, 10 ml are evaporated off. The

residue is taken up with nitric acid, c (HNO_3) = 0.5 mol/l, and made up to the mark in a 50 ml volumetric flask.

Wet Ashing for Carbon Determination (See COD Determination, Section 6.1.7)

2 g air-dried fine soil (for peat 0.5 g) are treated with 40 ml conc. sulfuric acid in a 250 ml volumetric flask. After 10 min 25 ml potassium dichromate solution, β ($K_2Cr_2O_7$) = 98.07 g/l, are added with cooling. The flask is held for 3 h at 120 °C in a drying oven with repeated swirling. The mixture is cooled, made up to 250 ml with water and an aliquot of 25 ml is back-titrated as described in Section 6.1.7 to determine the unreacted potassium dichromate.

To see the end point more readily 5 ml of a special acid mixture are added before the titration. (150 ml conc. H_2SO_4 + 150 ml conc. H_3PO_4 + 5 g $FeCl_3 \cdot 6H_2O$ are mixed and added to water with cooling. When cold the solution is made up to 1 l).

Digestion for Nitrogen Determination (See Kjeldahl Nitrogen, Section 6.1.20)

1 to 5 g air-dried fine soil are treated with a spatula tip of the selenium reaction mixture and 6 ml conc. sulfuric acid in a Kjeldahl flask. The sample is then heated until the residue is colourless. After cooling the solution is transferred to a 1 l round-bottomed flask which is then connected to a distillation apparatus. 25 ml sodium hydroxide, w (NaOH) = 30%, are added. Distillation is then carried out as described under the determination of Kjeldahl nitrogen.

6.3.1.5   Extraction

Extractable Ca, K, PO₄, SO₄

Calcium, potassium and phosphate are extracted with ammonium lactate – acetic acid in an equilibrium procedure:

5 g air-dried fine soil are shaken for 4 h with 100 ml of an extraction solution (consisting of 9 g lactic acid + 19 g acetic acid + 7.7 g ammonium acetate, made up to 1 l). The mixture is then filtered through a filter paper.

Sulfate is extracted with a sodium chloride solution in an equilibrium process:

50 g air-dried fine soil are shaken for 1 h with 250 ml sodium chloride solution, w (NaCl) = 1%. 3 g activated charcoal powder are then added, shaking is repeated briefly, and the mixture filtered.

Water-soluble B, Cl, SO₄, NO₃, Na, Ca, Mg

Boron is extracted with hot water:

25 g air-dried fine soil are boiled for 5 min with 50 ml water in a flask (low boron glass) and then filtered.

For the other groups the extraction is carried out as follows:

25 g air-dried fine soil are shaken with 125 ml water for 1 h and then filtered. This solution can then be used for further determinations. If all the water-soluble salts are to be determined together, 50 ml of the filtrate is placed in a weighed beaker and evaporated on a sand bath after addition of 5 ml hydrogen peroxide, w (H_2O_2) = 30%. The increase in weight in mg, divided by 10, gives the percentage of water-soluble salts.

6.3.2 Measurements

6.3.2.1 Water Binding and Capillary Pressure

Bound water found in soil is particularly important for the growth of plants and soil melioration. Bound water is divided into adsorption water and capillary water. Adsorption water is that which is surrounded by the soil particles and is not available to plants. Capillary water is that which is present in soil pores, has a surface meniscus and is subject to the laws of capillary action. The intensity of water binding is known as water tension and has to be overcome, for example, in dehydration. In the natural environment the water binding intensity is important for assessing the water supply in the soil or the lack of water available to plants. It is given in hPa or in cm of a water column, but mostly as its logarithm (= pF value). The possible water tension pF present in soil ranges from $-\infty$ to $+7$. A water tension of a 1000 cm water column corresponds to a pF value of 3. The permanent withering point of many plants is at pF 4.2.

Procedures for measuring the water capacity according to Richards, the hygroscopicity according to Mitscherlich and a simple method for determining the capillary pressure are described below.

Apparatus
Stainless steel 100 cm^3 cylinder for the determination of bulk density
Water bath
Reduced pressure vessel with suction pump

Reagents and Solutions
Sodium sulfate solution: 65 g sodium sulfate ($Na_2SO_4 \cdot 10H_2O$) are dissolved in 100 ml water.

Measurements
Water saturation:
Several soil samples in the steel cylinder are covered at the bottom with coarse filter paper and placed in a water bath in such a way that the water surface is first somewhat below and then a few mm above the upper edge of the cylinder. Water saturation takes place after a minimum of 5 h.
Water capacity:
The saturated samples are covered to protect against evaporation, placed together with the filter papers in a reduced pressure vessel and a pressure reduction of 60 hPa is set up (manometer). When no more water comes out, the cylinders are weighed without the filters. They are then dried to constant weight at 105 °C in a drying oven and reweighed after cooling. The dry contents of the cylinders are finally weighed without the cylinders.
Hygroscopicity:
20 g air-dried fine soil are slightly moistened with water in a petri dish (sprayer) and are then left in a desiccator over the sodium sulfate solution at 25 °C in vacuo. Control weighings are carried out after 4 days and then daily. When constant weight is reached the sample is dried at 105 °C to constant weight and the weight noted.

Calculation of Results

The water capacity WC per 100 cm^3 of the steel cylinder, measured in vol%, is calculated as follows:

$$WC = G_M - G_D$$

G_M moist weight of the steel cylinder, g
G_D dry weight of the steel cylinder, g

Calculation of the hygroscopicity HYG:

$$HYG = (Hy_m - Hy_d) \cdot 100/ Hy_d$$

HYG hygroscopicity, in % air-dried fine soil
Hy_m moist weight of the sample, g
Hy_d dry weight of the sample, g

Multiplication of HYG by 1.5 gives the approximate pore space of the soil at a permanent withering point of pF 4.2.

Simple Method for Determination of the Water Tension

Apparatus

Filter paper, e.g. Schleicher & Schüll no. 589 (white band)
Plastic sample boxes, ca. 100 ml
Room which can be air-conditioned, 20 °C

Reagents and Solutions

Pentachlorophenol solution: 0.5 g pentachlorophenol are dissolved in 100 ml methanol.

Measurement

10 to 20 g of the soil sample are placed in a box and an accurately weighed filter paper, which has previously been impregnated with pentachlorophenol solution and dried, is placed on top. The box is closed and made airtight with plastic insulating tape and is kept at 20 °C in an air-conditioned room for 1 week to set up an equilibrium. The filter paper is then rapidly removed and weighed immediately.

Calculation of Results

From the regression plots

$$pF = \log cm\ WCM = 6.24617 - 0.0723 \cdot M \text{ (for } M < 54\%)$$
$$pF = \log cm\ WCM = 2.8948 - 0.01025 \cdot M \text{ (for } M > 54\%)$$

the water tension pF, expressed in log cm water column, WCM, can be calculated. *M* is the moisture content of the filter paper given as a percentage of its dry weight. On use of a filter paper other than that given, another regression plot can be determined by exact pF measurements and parallel measurements in the way described above.

6.3.2.2   Particle Size

The particle size analysis should determine the ratios of the quantities of particles with different sizes, as the particle size strongly affects the physical properties of the soil.

Apparatus
Sieve set (pore sizes 2 mm, 0.63 mm, 0.2 mm, 0.063 mm)
100 ml measuring cylinder
Weighing bottles

Reagents and Solutions
Sodium pyrophosphate solution: 100 g sodium pyrophosphate ($Na_4P_2O_7$) are made up to 1 l
                                  with water.

Sample Preparation
The dried and coarsely ground crude soil is placed on a 2 mm sieve, the roots and stones being weighed separately. The fine soil (< 2 mm) is used in the analysis.

Measurement
20 g air-dried fine soil is allowed to soften with 25 ml sodium pyrophosphate solution for ca. 8 h and is then shaken for 1 h after addition of 200 ml water. The suspension is then poured on to a sieve set so that the filtrate runs directly into a 1 l measuring cylinder. The individual sieves are washed until the finer particles have passed through the corresponding sieve. The whole sieve set is then dried at 105 °C in a drying oven and each individual sieve is then weighed.

The suspension in the measuring cylinder is filled up to the 1000 ml mark and then shaken vigorously. After 9.5 min 10 ml are removed with a pipette immersed to 20 cm below the water surface and this volume is transferred to a weighing bottle. The fraction < 20 μm is present in the suspension. The fraction < 10 μm is removed after 18.5 min by pipetting 10 ml suspension from 10 cm below the water surface and is also transferred to a weighing bottle. Finally the clay fraction < 2 μm is taken from a depth of 4 cm after 3 h and 5 min.

All the samples taken are dried at 105 °C in the drying oven and are then weighed. 25 mg sodium pyrophosphate are subtracted from the quantity of substance determined.

Interfering Factors
The method is based on the rate of sinking of spherical particles in water. The more the particles deviate from the ideal sphere, the lower their sinking rate is, i.e. the particles are assigned to a coarser particle size fraction than that to which their average diameter corresponds.

Calculation of Results

The weighings of the individual fractions can be based on the crude soil sample or the proportion of fine soil, depending on the requirements. The weights of the evaporated samples determined in the weighing bottles (from which the weight of sodium pyrophosphate has been subtracted) are multiplied by 5. This gives the proportion of the corresponding fraction as a percentage of the original quantity weighed out.

6.3.2.3   Hydrolytic Acidity (H-value)

Determination of the total acid content of soil involves determination of the free H_3O^+ ion concentration as well as that of complex bound H_3O^+. Indirect methods are mainly employed in which hydrolysable salts are brought into contact with the soil and the free acid formed by reaction of absorbed H_3O^+ ions with the anion is measured by titration. The value obtained is multiplied by empirical factors to estimate the total acid.

Apparatus

Titration equipment

Reagents and Solutions

Calcium acetate solution: 88.09 g calcium acetate $((CH_3COO)_2Ca \cdot H_2O)$ are made up to 1 l with water. The colour of the solution is adjusted to pink with sodium hydroxide, c (NaOH) = 0.1 mol/l, and a few drops of phenolphthalein.

Measurement

100 g air-dried fine soil are treated with 250 ml calcium acetate solution and shaken for 1 h. After filtration the first 30 ml are discarded and 125 ml of the filtrate are titrated against phenolphthalein until a pink colour appears.

Calculation of Results

The sodium hydroxide consumed is multiplied by empirical factors to give values for the total soil acidity:

for adjustment to pH 7	$F = 1.5$,
for adjustment to pH 7.5	$F = 2.0$,
for adjustment to pH 8.0	$F = 2.5$,
for adjustment to pH 8.5	$F = 3.25$.

If the titration is performed using phenolphthalein F = 3.25 is used.

The hydrolytic acidity is given in ml sodium hydroxide, c (NaOH) = 0.1 mol/l, per 100 g soil. The total acidity in ml sodium hydroxide, c (NaOH) = 0.1 mol/l, is calculated as follows:

Total acidity $= 2 \cdot x \cdot 3.25$

x consumption of sodium hydroxide, c (NaOH) $= 0.1$ mol/l, ml

The sum of total acid and base saturation (S-value, see Section 6.3.2.4) gives the total base saturation (exchange capacity) in mmol/100 g soil. Earlier data given in the unit Val can be converted into the unit mol with the same numbers if the equivalent weight is given.

6.3.2.4   Exchangeable Basic Substances (S-value)

In determining the S-value the cations, such as calcium, magnesium, sodium, potassium and ammonium, which are fixed to negatively charged soil colloids are covered, but not carbonates. The S-value is important for estimating the potential delivery of bases and therefore the soil fertility.

Apparatus
Titration equipment

Reagents and Solutions
Hydrochloric acid, c (HCl) $= 0.1$ mol/l
Sodium hydroxide, c (NaOH) $= 0.1$ mol/l
Sodium potassium tartrate solution, w (SPT) $= 10\%$
Ethanol – water mixture (60:40 v/v)

Sample Preparation
15 to 25 g moist fine soil are placed on a fluted filter paper and the alcohol – water mixture poured over it several times to displace the soil water without causing exchange processes to take place. The residue is dried with the filter paper at 105 °C in a drying oven.

Measurement
Depending on the calcium carbonate content of the soil, between 1 and 10 g of fine soil are weighed out (1 g at w ($CaCO_3$) $= 30$ to 50%, 10 g at w ($CaCO_3$) $= 5\%$), treated with 100 ml hydrochloric acid, c (HCl) $= 0.1$ mol/l, and shaken for 1 h. After filtration 20 ml of the filtrate are treated with 5 ml sodium potassium tartrate solution, w (SPT) $= 10\%$, and the solution titrated against sodium hydroxide, c (NaOH) $= 0.1$ mol/l, with phenolphthalein until a pink colour appears. The blank consists of 20 ml hydrochloric acid, c (HCl) $= 0.1$ mol/l, and 5 ml sodium potassium tartrate solution, w (SPT) $= 10\%$, titrated in the same way.

Calculation of Results
The difference between the titration values for the blank and the soil extract is multiplied by a factor 50 for an initial weight of soil of 1 g (factor 5 for an initial weight of 10 g), thus giving the S-value in mol/100 g soil. This value includes carbonate so a carbonate determination must be carried out in parallel. The result of the S-value titration is then corrected for the carbonate content of the soil.

6.3.2.5   Cation Exchange Capacity

The exchange capacity of soil is a measure of the proportion of colloidal substances whose surfaces can function as cation exchangers. In the determination the fixed cations are replaced by a high concentration of readily exchangeable ones. The quantity of exchanged cations corresponds to the exchange capacity of the soil. The values for mineral soils lie between 15 and 40 mol/100 g soil and for soils high in humus up to 300 mol/100 g soil.

Apparatus
Distillation equipment as described in Section 6.1.20

Reagents and Solutions
Ammonium oxalate solution, c (NH_4-ox) = 0.2 mol/l
Calcium carbonate, powdered
Activated charcoal, powdered

Measurement
250 ml ammonium oxalate solution, c (NH_4-ox) = 0.2 mol/l, are added to 20 g air-dried fine soil together with 5 g activated charcoal (for fixation of ammonium humates) and 0.5 g calcium carbonate (as buffer). The mixture is then shaken for 2 h and filtered. 2 drops of conc. sulfuric acid are added to the filtrate. The ammonium content of 25 ml of this solution is then measured after distillation as described in Section 6.1.20. A blank sample, consisting of 25 ml ammonium oxalate solution, c (NH_4-ox) = 0.2 mol/l, is treated in the same way.

Calculation of Results
The difference between the ammonium content of the blank and that of the soil extract gives the quantity of NH_4^+ ions taken up by the soil. The exchange capacity is given in mol/100 g soil.

6.3.2.6   Carbonate Content

Knowledge of the calcium carbonate content of a soil is of considerable importance as this parameter influences, for example, the soil texture and permeability. Also, chemical processes in the soil are affected by the carbonate content.

The volumetric method of determination is described below.

Apparatus
Measuring apparatus as shown in Fig. 40

Reagents and Solutions
Hydrochloric acid, w (HCl) = 10%
Potassium chloride solution, w (KCl) = 2%

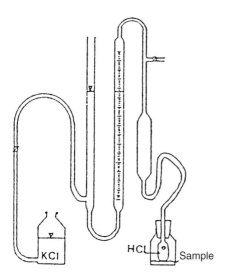

Fig. 40: Instrument for the determination of carbonate in soil

Measurement

Depending on the results of a qualitative preliminary test with hydrochloric acid, between 2 and 10 g of air-dried fine soil are placed in the gas generator and the insert filled with 20 ml hydrochloric acid, w (HCl) = 10%. After connecting to the apparatus, the graduated tube is filled by raising the level container. The gas generator is then tilted so that the hydrochloric acid comes into contact with the soil. Depending on the level set in the graduated tube, pressure compensation is attained by raising or lowering the level container. After ca. 10 min the gas volume is read off.

Interfering Factors

The measurement is relatively imprecise for a carbonate content of less than w (CO_3^{2-}) = 1%. If magnesium and iron carbonates are present, lower values are obtained owing to the lower reactivity of both compounds and to the altered stoichiometric factors.

Calculation of Results

The proportion by weight of carbonate in the soil, based on calcium carbonate, is calculated as follows and is given as a percentage:

$$w(CaCO_3) = \frac{V \cdot P \cdot 0.12}{(273 + t) \cdot W}$$

V volume of CO_2 measured, ml
P air pressure, hPa $\approx$ mbar
t room temperature, °C
W weight of soil, g

6.3.2.7   Humic Substances

The organic material in soil consists to a considerable extent of humic substances. Because of their large specific surface area, these are very important for ion exchange, water and nutrient binding and buffering capacity. Humic substances do not have the same composition, but consist mostly of cyclic building blocks with reactive carboxyl, carbonyl and hydroxyl groups which can bond to medium and high molecular weight substances. As it is impossible to characterise humic substances exactly, the various groups are classified conventionally according to their extractability with bases and their insolubility in acids. Fulvic and humic acids can be extracted with base, but not humins. Humic acids can be precipitated with acid, but fulvic acids cannot.

Apparatus
Shaker
Spectrophotometer or fixed filter photometer with 472 and 664 nm filters
Membrane filtration apparatus with 0.45 μm filters

Reagents and Solutions
Sodium hydroxide, c (NaOH) = 0.1 mol/l
Sodium pyrophosphate solution, c ($Na_4P_2O_7$) = 0.1 mol/l
Hydrochloric acid, w (HCl) = 36%
Humic acid (e.g. Merck)

Sample Preparation
20 to 50 g air-dried fine soil (the absolute quantity of organic material should be between 0.05 and 0.5 g) are shaken with 200 to 500 ml of a mixture of sodium hydroxide and sodium pyrophospha-te (1:1) for 5 h. The mixture is then centrifuged for ca. 10 min at ca. 3000 min^{-1} and finally filtered through a membrane.

Calibration and Measurement
The extinction of the filtered solution is determined photometrically at 472 and 664 nm. The concentration of humic substances is taken from a calibration curve, which is plotted using a commercially available humic acid at the given wavelengths. To determine the proportion of humic acid, the pH of the filtered solution is adjusted to < 1 with hydrochloric acid. After allowing the mixture to settle overnight, the liquid is decanted off and the residue centrifuged. After rinsing several times with acidified water, the salts are removed and the residue is dried in a weighing bottle at 105 °C and weighed.

The extinctions of the supernatant after acid precipitation (fulvic acids) and of the humic acids in the precipitate redissolved in sodium hydroxide can be measured photometrically at 472 and 664 nm and compared with the commercial humic acid.

Interfering Factors
Besides humic substances, other soil components can give rise to a yellow-brown colour. In addition the commercial humic acids used as a reference have a different extinction (often only slight-

ly different to that of the extracted substances at the same concentration). This is because the molecular weights and molecular structures are seldom the same. The error thus caused must, however, be accepted.

Calculation of Results

The content of humic substances is given in mg/g fine soil. The higher the proportion of humic substances in the organic material of the soil is (determined, for example, as the loss on ignition), the greater is the extent of humification. The quotient of the extinctions at 472 and 664 nm is usually given as the $Q_{4/6}$ value. Generally the molecular size and the content of carbon and nitrogen increase as the quotient decreases. At a quotient of < 3 high molecular weight grey humic acids dominate, at 4 to 5 brown humic acids and at > 5 low molecular weight fulvic acids.

6.3.2.8   SAR Value (Sodium Adsorption Ratio)

At a high sodium content in soil in relation to other cations (especially calcium and magnesium) there is a danger of oversalting and thus of lower agricultural profitability. The determination of the SAR value can be used to characterise the problems caused by sodium. The SAR value of irrigation water besides that of soil can also be important.

A nomogram (Fig. 41) simplifies the determination of the SAR value. The measured sodium concentration on ordinate I is connected with the value for the sum of the calcium and magnesium concentrations on ordinate II and the SAR value is read off from the diagonal.

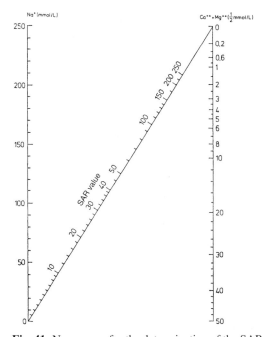

Fig. 41: Nomogram for the determination of the SAR value in soil extracts

Figure 42 shows a nomogram for sprinkler water. For the calculation of the SAR value of irrigation water see Section 7.5.

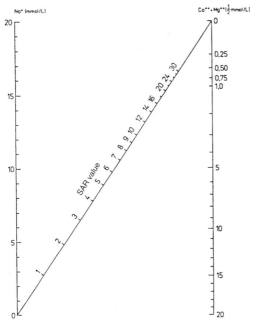

Fig. 42: Nomogram for the determination of the SAR value in sprinkler water

Measurement

A soil paste is prepared by adding sufficient water to ca. 250 g moist or air-dried soil with stirring so that it is saturated without the presence of free water. After leaving for 2 h, the paste is transferred to a suction filter and the water drawn off. The parameters sodium, calcium and magnesium are measured in the soil solution.

Calculation of Results

The SAR value can be calculated as follows (the concentration is given in milliequivalents (mmol)):

$$SAR = \frac{Na^+}{\sqrt{(Ca^{2+} + Mg^{2+})/2}}$$

7 Interpretation of Test Results

The evaluation and interpretation of test results on water, waste water and soil requires considerable practical experience. International and national guidelines and threshold values can be of great assistance. These are based on experience in the areas of human and ecotoxicology, nutrition, agriculture and technology. Simple acceptance of recommended and threshold values is, however, not advisable. In an individual case the local conditions, the variance in sampling and analysis procedures and the requirements for use should be taken into consideration when evaluating data.

The guidelines, recommendations and tables of threshold values given below are only a selection and it cannot be claimed that the lists are exhaustive.

7.1 Ground Water

The interpretation of chemical and bacteriological results of tests on ground water must take the background values into account as far as possible. The distinction between natural and man-made contamination is not always easy to make, as some substances can get into ground water both naturally and as a result of human activity. Comparative measurements therefore have to be relied upon. Temporal variations in the concentration must also be considered. As both ground water and surface water are used as a source of drinking water or for irrigation, the recommended values for these water uses should be used in the evaluation. In addition the EEC Guideline of 1980 concerning the protection of ground water against contamination by certain specific hazardous substances should be referred to. The particularly hazardous substances named in list I of the appendix to the Guideline (e.g. cyanides, mercury compounds, organochlorine, -phosphorus and -tin compounds) may no longer be discharged, while those in list II (metals and metalloids) may only be discharged into ground water after extensive testing.

7.2 Surface Water

For surface water various guidelines are in force, e.g. the German requirements for the minimum quality for flowing water in relation to the water classes II/III (Table 29), the German requirements for fishery waters (Table 30) or the EEC Bathing Water Quality Directive 76/160/EEC (Table 31). The parameters listed and their threshold values provide a framework based on the relationship between quality class and self purification and which guarantees a minimum quality. This also applies when a body of water or part of it is used to carry waste water. In individual cases the minimum quality requirements can be made stricter by the state authorities responsible or other parameters can be added to the list.

Table 29: Minimum requirements for flowing water systems in Germany
(Basis: German water classes II/III) (mg/l)

Parameter	Minimum requirement	1. Purification necessary
Maximum temperature (°C)		
a) Cool summer waters	25	25
b) Warm summer waters	28	28
O_2	4	4
pH value	6–9	6–9
NH_4^+-N	1	2
BOD_5 (without nitrification inhibition)	7	10
COD	20	30
P	0.4	1
Fe	2	3
Zn	1	1.5
Cu	0.05	0.06
Cr	0.07	0.1
Ni	0.05	0.07

Table 30: Quality requirements for fishery waters in Germany
(Basis: German water class II) (mg/l)

Parameter	Salmonide waters	Cyprinide waters
Maximum temperature (°C)		
a) Cool summer waters	20	25
b) Warm summer waters	20	28
O_2	6	4
pH	6.5–8.5	6.5–8.5
NH_4^+-N	1	1
BOD_5 (without nitrification inhibition)	6	6
COD	20	20
Fe	2	2
Zn		
a) at β(Ca) = 4 mg/l	0.03	0.3
b) at β(Ca) = 20 mg/l	0.2	0.7
c) at β(Ca) = 40 mg/l	0.3	1
Cu, dissolved		
a) at β(Ca) = 4 mg/l	0.005	0.005

Continuation **Table 30:**

Parameter	Salmonide waters	Cyprinide waters
b) at β (Ca) = 20 mg/l	0.022	0.022
c) at β (Ca) = 40 mg/l	0.04	0.04
NO_2^--N	0.015	0.015

Table 31: EEC Bathing Water Quality Directive 76/160/EEC 1975

Parameter	Recommended value	Compulsory limit
Microbiological parameters		
Total coliform/100 ml	500	10000
Faecal coliform/100 ml	100	2000
Faecal streptococcus/100 ml	100	–
Salmonellae/l	–	0
Intestinal virus PFU/10 l	–	0
Physico-chemical parameters		
pH	–	6–9
Transparency (m)	2	1
Colour	–	No abnormal change
Mineral oils mg/l	– (< 0.3)	No visible film
Anionic surfactants mg/l	– (< 0.3)	No foam formation
Phenol index mg/l	0.005	0.005
Tar residues mg/l	None	–

7.3   Drinking Water

The quality of drinking water is regulated in most countries by recommendations or legal requirements. Of universal importance are the recommendations of the World Health Organisation (WHO)
of 1993 (Table 32). Furthermore the EC Directive on the Quality of Water intended for Human Consumption 98/83/EC (Table 33) is to be assimilated into the national regulations of EC countries.
The threshold values of the German ordinance on drinking water of 1991 are given in Table 34.

If surface water is to be used for preparing drinking water, the recommended values of the EEC
Guideline of 1975 (Table 35) should be consulted. The German ordinance on drinking water gives
threshold values for additives used in drinking water treatment, besides hygienic threshold values
(Table 36).

Table 32: WHO recommendations for drinking water quality, 1993 (extract)

I Microbiological parameters

Organism	Recommended value (count/100 ml)	Remarks
Drinking water		
E. coli or		
thermoltolerant coliforms	0	
Treated water on introduction into the water system		
E. coli or		
thermoltolerant coliforms	0	
Total coliforms	0	
Treated water in the mains		
E. coli or		
thermoltolerant coliforms	0	
Total coliforms	0	For large water treatment works and for large numbers of samples total coliforms must not be present in 95% of the samples over a 12 month period.

II Chemical parameters

Parameter	Recommended value	Remarks
Inorganic substances (mg/l)		
As	0.01	Provisional value
B	0.3	
Ba	0.7	
Cd	0.003	
CN	0.07	
Cr	0.05	Provisional value
Cu	2	At ≥ 2 possibly detrimental to taste
F	1.5	Climatic conditions, quantities of water collected, and other feeding habits should be taken into consideration
Hg	0.001	
Mn	0.5	Provisional value
Mo	0.07	
Ni	0.02	
NO_3	50	
NO_2	3	Provisional value

Continuation **Table 32:**

Parameter	Recommended value	Remarks
Pb	0.01	Value cannot be adhered to immediately everywhere; all measures for immediate reduction are to be taken.
Sb	0.005	Provisional value
Se	0.01	
Organic substances (µg/l)		
Aldrin, dieldrin	0.03	
Benzene	10	
Benzo(a)pyrene	0.7	
Chlordane	0.2	
Chloroform	200	
2,4-DB	90	
DDT	2	
Dichloromethane	20	
Di-2-ethylhexyl phthalate	8	
1,2-Dichloroethane	30	
Nitrilo-triacetic acid	200	
Formaldehyde	900	
Heptachlor and heptachlor epoxide	0.03	
Lindane	2	
MCPA	2	
Methoxychlor	20	
Monochlorobenzene	300	
Pentachlorophenol	9	Provisional value
Styrene	20	
Tetrachloroethylene	40	
Carbon tetrachloride	2	
2,4,6-Trichlorophenol	200	
Toluene	700	
Trichloroethylene	70	Provisional value
1,1,1-Trichloroethane	2000	Provisional value
Vinyl chloride	5	
Xylenes	5	

The EC Directive on the Quality of Water intended for Human Consumption 98/83/EC lists in the appendix values for the microbiological, chemical and so-called indicator parameters. The threshold values (minimum requirements) laid down in parts A and B are to be adhered to, while the parameters given in part C are to be used only for monitoring purposes and adherence to the obligations of EC member countries with respect to eventual reproducibility.

Table 33: Council Directive of 1998 on the quality of water intended for human comsumption (extract)

Parameter	Parametric value	Notes
Part A: Microbiological parameters		
E. coli	0/100 ml	
Enterococci	0/100 ml	
(The following applies to water offered for sale in bottles or containers)		
E. coli	0/250 ml	
Enterococci	0/250 ml	
Pseudomonas aeruginosa	0/250 ml	
Colony count 22 °C	100/ml	
Colony count 37 °C	20/ml	
Part B: Chemical parameters		
Acrylamide	0,10 µg/l	monomer
Antimony	5,0 µg/l	
Arsenic	10 µg/l	
Benzene	1,0 µg/l	
Benzo(a)pyrene	0,010 µg/l	
Boron	1,0 mg/l	
Bromate	10 µg/l	
Cadmium	5,0 µg/l	
Chromium	50 µg/l	
Copper	2,0 mg/l	at the tap
Cyanide	50 µg/l	
1,2-dichlorethane	3,0 µg/l	
Epichlorohydrin	0,10 µg/l	monomer
Fluroride	1,5 mg/l	
Lead	10 µg/l	at the tap
Mercury	1,0 µg/l	
Nickel	20 µg/l	at the tap
Nitrate	50 mg/l	
Nitrite	0,50 mg/l	
Pesticides	0,10 µg/l	applies to each individual pesticide
Pesticides – Total	0,50 µg/l	

Table 33:   (Continued)

Parameter	Parametric value	Notes
Polycyclic aromatic	0,10 µg/l	sum of concentrations of 4 reference substances
Selenium	10 µg/l	
Tri- and tetrachloroethene	10 µg/l	sum of concentration
Tri- and tetrachloroethene	10 µg/l	sum of concentration
Trihalomethane – Total	100 µg/l	sum of concentration of 4 reference substances
Vinyl chloride	0,50 µg/l	
Part C: Indicator parameters		
Aluminium	200 µg/l	
Ammonium	0,50 mg/l	
Chloride	250 mg/l	the water should not be aggressive
Clostridium perfringens	0/100 ml	if the water is influenced by surface water
Colour	acceptable to consumers	
Conductivity	2500 µS/cm (20 °C)	the water should not be aggressive
Hydrogen ion concentration	$\geq 6,5$ and $\leq 9,5$	the water should not be aggressive, for bottled water reduced to pH 4,5
Iron	200 µg/l	
Manganese	50 µg/l	
Odour	acceptable to consumers	
Oxidisability	5,0 mg/l O_2	needs not be measured if TOC is analysed
Sulphate	250 mg/l	
Sodium	200 mg/l	
Taste	acceptable to consumers	
Colony count 22 °C	no abnormal change	
Coliform bacteria	0/100 ml	for bottled water 0/250 ml
Total organic carbon (TOC)	no abnormal change	no need for less than $10\,000$ m^3/d
Turbidity	acceptable to consumers	
Tritium	100 Bq/l	
Total indicative dose	0,10 mSv/year	

Table 34 gives the threshold values of the German ordinance on drinking water (Trinkwasser-verordnung, TVO) of 1991. These values also apply to water used in the food industry. Microbiological measurements and measurements according to Appendix 2 of the TVO are obligatory, while the parameters listed in Appendix 4 are only to be tested if required by the authorities responsible.

Table 34: Threshold values given in the German ordinance on drinking water (TVO)

Parameter	Threshold value
Threshold values for chemical substances (Appendix 2 TVO) (mg/l)	
As	0.04
Pb	0.04
Cd	0.005
Cr	0.05
CN	0.05
F	1.5
Ni	0.05
NO_3	50
NO_2	0.1
Hg	0.001
Polycyclic aromatic hydrocarbons (six substances as C)	0.0002
Organochlorine compounds as Σ of 1,1,1-trichloroethane, trichloroethylene, tetrachloroethylene, and dichloromethane	0.01
Carbon tetrachloride	0.003
When required	
Plant protection agents (individual substances)	0.0001
Plant protection agents (total)	0.0005
Sb	0.01
Se	0.01
Microbiological parameters	
Similar threshold values to those given in Table 33. In addition the colony count in disinfected drinking water should not exceed the recommended value of 20 per ml at 20 °C.	

Table 34: (Continued)

Parameter	Threshold value
Parameters and threshold values (Appendix 4 TVO) (mg/l)	
Colour (spectral absorption coefficient Hg 436 nm), m^{-1}	0.5
Turbidity (formazine units)	1.5
Threshold level for odour	2 at 12 °C
	3 at 25 °C
Temperature	25 °C (not warmed drinking water)
pH	6.5–9.5 (see TVO)
Electrical conductivity, μS cm^{-1}	2000
Oxidisability, mg/l	5
Threshold values for chemical substances (Appendix 4 TVO) (mg/l)	
Al	0.2
NH$_4$	0.5 (up to 30 if caused by geological conditions)
Ba	1
Ca	400
Cl	250
Fe	0.2
K	12 (up to 50 if caused by geological conditions)
Kjeldahl-N	1
Mg	50 (up to 120 if caused by geological conditions)
Mn	0.05
Na	150
Phenols	0.0005
PO$_4$	6.7
Ag	0.01
SO$_4$	240 (up to 500 if caused by geological conditions)
Hydrocarbons	0.01
Substances extractable with CHCl$_3$	1
Anionic surfactants, nonionic surfactants	0.2

Surface water is being increasingly used as a source of drinking water. Since here man-made contamination is usually higher than in the case of ground water, the EEC published a guideline in 1975 summarising the quality requirements for surface water (Table 35). Only the obligatory pre-scribed values are given; the recommended values have been omitted.

Table 35: EEC Guideline on quality requirements of surface water for drinking water preparation (mg/l) (extract)

Parameter	Treatment categories		
	A1	A2	A3
Colour after single filtration, Pt colour index	20	100	200
Temperature (°C)	25	25	25
NH_4	–	1.5	4
NO_3	50	50	50
F	1.5	–	–
Fe	0.3	2	–
Cu	0.05	–	–
Zn	3	5	5
As	0.05	0.05	0.1
Cd	0.005	0.005	0.005
Cr, total	0.05	0.05	0.05
Pb	0.05	0.05	0.05
Se	0.01	0.01	0.01
Hg	0.001	0.001	0.001
Ba	0.01	1	1
CN	0.05	0.05	0.05
SO_4	250	250	250
Phenols	0.001	0.005	0.1
Dissolved or emulsified hydrocarbons	0.05	0.2	1
Polycyclic aromatic hydrocarbons	0.0002	0.0002	0.001
Total pesticides	0.001	0.0025	0.005

Category A1:
simple physical treatment and disinfection (e.g. rapid filtration and disinfection)
Category A2:
normal physical treatment, chemical treatment and disinfection (e.g. prechlorination, coagulation, flocculation, decantation, filtration and disinfection)
Category A3:
intensive physical and chemical treatment (e.g. break-point chlorination, coagulation, flocculation, decantation, filtration, adsorption on activated charcoal, disinfection (ozone, chlorine)).

Table 36 gives a list of additives which may be used in the preparation of drinking water in Germany according to Appendix 3 of the TVO of 1991.

Table 36: Additives for drinking water treatment (mg/l) (extract)

Substance	Maximum permitted addition	Maximum permitted level in treated drinking water
Chlorine, sodium, calcium and magnesium hypochlorites, chlorite of lime	1.2	0.3
Chlorine dioxide	0.4	0.2
Ozone		
a) Disinfection	10	0.05
b) Oxidation	–	0.01
Silver, silver chloride, silver sulfate, Sodium – silver chloride complex	–	0.08
Hydrogen peroxide, sodium peroxodisulfate, potassium monopersulfate	17	0.1
Sulfur dioxide, sodium sulfite, calcium sulfite	5	2
Sodium thiosulfate	6.7	2.8
Sodium silicate, sodium hydroxide, sodium carbonate, sodium hydrogen carbonate	–	40

Besides the substances listed, the following may be used in the processing of drinking water: calcium carbonate, partially calcined dolomite, calcium oxide, calcium hydroxide, magnesium carbonate, magnesium oxide, magnesium hydroxide, sodium carbonate, sodium hydroxide, sodium hydrogen carbonate, sulfuric acid, hydrochloric acid, and sodium, potassium and calcium phosphates.

The assessment of the corrosive properties of water is particularly important for its use in contact with various materials. Table 37 lists the requirements given in DIN 50 930.

7.4   Water for Use in Construction

Water destined for building purposes should satisfy certain requirements in order to avoid damage to building materials. Basically the water should be as pure as possible (e.g. drinking water quality) as impure or very salty water can lead to damage, such as a decrease in the hardness of concrete. The presence of mineral acids, humic acids and carbonic acid can retard the hardening of low calcium cements by reacting with the calcium carbonate before setting begins. Oils and fats can coat the reactive surfaces of the cement components thus preventing the entry of water essential for hardening. Increased concentrations of dissolved organic substances can delay hardening in the same way. The following types of water may not be used to mix concrete:

– sea water with more than 3.5% salt,
– water with more than 3.5% dissolved sulfate,
– organically polluted waste water,
– water with pH < 4 (may possibly be neutralised before use).

Table 37: Requirements concerning the corrosion properties of water according to DIN 50 930

Parameter	Unalloyed and low alloy iron materials	Hot-galvanised iron materials	Nonrusting steels	Copper and copper alloys
Oxygen (mg/l)	> 3 mg/l		As low as possible	As low as possible
pH	As high as possible, but < 8.5	>7.5	< equilibrium (< saturation pH)	As high as possible
Acidic capacity $K_{S\,4.3}$ (mmol/l)	> 2 mmol/l (not < 1.5)	> 1 mmol/l (better < 2)		
Calcium (mg/l)	> 0.5 mmol/l	> 0.5 mmol/l		
Other criteria (in mmol/l)	$\dfrac{Cl^- + 2\,SO_4^{2-}}{K_{S4,3}} > 1$ $\dfrac{Cl^- + 1/2\,SO_4^{2-}}{NO_3^-} > 2$	$\dfrac{Cl^- + 1/2\,SO_4^{2-}}{K_{S4,3}} > 1$		$\dfrac{HCO_3^-}{SO_4^{2-}} > 2$

Table 38: Recommended values for mixing water for concrete (mg/l)

Component	Recommended value
pH	ca. 7
Free CO_2	25
Sulfide S	Not detectable
SO_4	250
Cl	1500
NH_4	100
Mg	200
$KMnO_4$ consumption	25
Humic acids and hydrocarbons	Not detectable
For iron reinforcement of concrete:	
Cl	100
NO_3	20 to 50

The recommended values given in Table 38 should be adhered to for the water used to mix highly stressed concrete, e.g. that used in foundations.

Hardened concrete can be corroded by contact with certain types of water. Threshold values for the assessment of such water, according to DIN 4030, are given in Table 39.

Table 39: Threshold values for assessment of waters aggressive to concrete according to DIN 4030 (mg/l)

Aggressive components	Degree of corrosion		
	Weak	**Severe**	**Very severe**
pH	6.5–5.5	5.5–4.5	> 4.5
Calcium aggressive CO_2	15–30	30–60	> 60
(marble test according to Heyer)			
NH_4	15–30	36–60	< 60
Mg	100–300	300–1500	> 1500
SO_4	200–600	600–3000	> 3000

7.5   Water for Irrigation

Guidelines concerning the quality of water for irrigation can only be usefully applied if climate, soil, plant types and the irrigation system used are taken into account.

Table 40 gives a rough classification of the salt contents of waters.

Table 40: Categories of water salt contents

Degree of salt content	Quantity of dissolved salts (g/l)
Weakly salty	< 0.15
Moderately salty	0.15–0.5
Strongly salty	0.5–1.5
Very strongly salty	1.5–3.5

Various cations and anions can adversely affect irrigation:

Magnesium
High concentrations can impair the growth of plants. According to the formula:

$$x = Mg^{2+} \cdot 100/Ca^{2+} + Mg^{2+}$$

a recommended value *x* can be calculated, given in 1/2 mmol/l. A value of 50 is harmful to many plants.

Carbonate/Bicarbonate

Waters containing carbonate are harmful to alkaline, clay or compact soils. However, they can be advantageous to acidic or sandy soils.

Chloride

A higher chloride concentration can be harmful to many cultivated plants, particularly fruit trees. Tolerance limits are known for many plant types (Table 41).

Table 41: Chloride tolerance limits for plants

Effect	Chloride concentration (mg/l)
Weak (suitable for almost all plants)	< 70
Moderate (suitable for chloride-tolerant plants, otherwise slight to moderate damage)	70–140
Medium (suitable for salt-resistant plants)	140–280
Strong (slight to moderate damage to salt-resistant plants)	> 280

Boron

At low concentrations boron is an important element for plant growth but can have toxic effects at higher concentrations. Recommended values are given in Table 42.

SAR Value

The SAR value (sodium adsorption ratio) is often employed in assessing the suitability of water for irrigation. The calculation is carried out according to the formula given in Section 6.3.2.8. Figure 43 shows an aid for the assessment of water.

Table 42: Boron tolerance limits for plants

Effect	Boron concentration (mg/l)
Weak (suitable for all plants)	0.3–1.0
Medium (suitable for boron-tolerant plants)	1.0–2.0
Strong (suitable for boron-resistant plants)	2.0–4.0

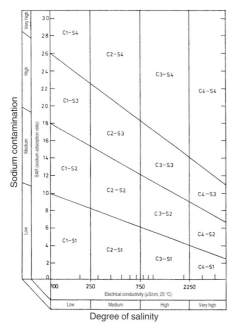

Fig. 43: Classification of irrigation waters (Richards 1969)

The symbols C_1 to C_4 and S_1 to S_4 in Fig. 43 have the following meaning:

C_1 water with low salt content (up to 0.15 g/l)
C_2 water with medium salt content (0.15 to 0.5 g/l)
C_3 water with high salt content (0.5 to 1.5 g/l)
C_4 water with very high salt content (1.5 to 3.5 g/l)
S_1 SAR < 10 for water with low salt content
       SAR < 2.5 for water with high salt content
S_2 SAR 10 to 18 for water with low salt content
       SAR 2.5 to 7 for water with high salt content
S_3 SAR 18 to 26 for water with low salt content
       SAR 7 to 11 for water with high salt content
S_4 SAR > 26 for water with low salt content
       SAR > 11 for water with high salt content

7.6   Waste Water

In order to determine threshold values for the discharge of waste water, a distinction must be made between direct and indirect discharge. Direct dischargers transfer treated waste water (untreated only in exceptional cases) directly into a body of water, while indirect dischargers transfer the water into the public drainage system and thereby usually to a sewage plant.

In general, the following safety targets are to be taken into consideration:

– for personnel: protection against H_2S, HCN, SO_2, CO_2, extreme pH values, high temperatures,
– for construction: protection against corrosion/damage caused by extreme pH values, sulfate, calcium aggressive CO_2, solid deposits,
– for the function of sewage plants: protection against reduced efficiency or breakdown caused by excessive contamination,
– for the quality of the receiving body of water: protection against undesired concentrations of nutrients and pollutants.

In the case of direct discharge of waste water threshold values can be set based on the principles of emission or pollution. Pollution threshold values are water course quality criteria which describe the condition of the water course. They take into account the total pollution and the self-purification capability (see Table 29, Section 7.2). This requirement is used, for example, in the EEC Guidelines concerning the quality requirements for surface water for the preparation of drinking water of 1975 (see Table 35) and in the EEC Bathing Water Quality Directive 76/160/EEC (see Table 31). Emission threshold values, on the other hand, give the discharge conditions for the maximum permissible concentration at the discharge point from a drain or sewage treatment plant. These rules have the advantage of being more easily monitored.

In Germany discharge conditions have been drawn up for domestic and industrial waste water as minimum requirements (emission thresholds) in the appendices to the general administrative framework regulation on minimum requirements for the discharges of waste water into waters (situation in 1996). In the meantime special regulations exist for over 50 areas of industry. The threshold values for some areas are shown in Table 43. They refer to a qualitative random sample and a 2-hourly pooled sample.

Table 43: Selected threshold values for the direct discharge of industrial waste water according to § 7a of the Federal Water Act (WHG) (mg/l)

Area of industry	COD	BOD$_5$	NH$_4$-N	Σ N	Total P	Hydrocarbons
Sugar production	250	25	10	30	2	
Potato processing	150	25	10		2	
Milk processing	110	25	10		2	
Breweries	110	25	10		2	
Fish processing	110	25	10	25	2	2
Alcohol production	110	25	10			
Leather production	250	25	10		2	
Stones and soil	80					
Crude oil processing	80	25		40	1.5	2

The values listed mostly require the application of generally recognised technical measures. In addition the minimum requirements according to the state of the art are partially in force for other parameters (in particular toxic substances) and other areas of industry. The details of the regulations can be found in specialist literature relating to water laws.

Waste water from waste dumps is becoming increasingly important. Since 1996 its discharge has been regulated by Appendix 51 of the general administrative framework regulation on minimum requirements for the discharges of waste water into waters. Here both the generally recognised technical measures and the state of the art are designated as requirements. It is important that for untreated seepage water with a COD of > 4000 mg/l a run-off rate is adhered to, which corresponds to a lowering of the COD of at least 95%. The threshold values given in Table 44 refer to a qualified random sample or a 2-hourly pooled sample.

Table 44: Requirements for the discharge of waste water from overground waste dumps

Parameter	Concentration (mg/l)
Generally recognised technical measures	
COD	200
BOD_5	20
ΣN	70
Total P	3
Hydrocarbons	10
NO_2-N	2
State of the art	
AOX (adsorbable organic halogen)	0.5
Hg	0.05
Cd	0.1
Cr	0.5
Cr(VI)	0.1
Ni	0.5
Pb	0.5
Cu	0.5
Zn	2
CN, easily released	0.2
Sulfide S	1
Toxicity towards fish (TF)	2

The amended German Wastewater Association (ATV) leaflet A 115 of 1994 gives recommendations for the drawing up of requirements for the discharge of industrial waste water into public sewage plants (Table 45). In certain German towns and villages, however, stricter requirements have been formulated. The recommended values according to A 115 for hazardous substances as understood by § 7a of the Federal Water Act (WHG) are only valid for discharges for which there are no requirements according to the state of the art in the appendices of the general administrative framework regulation on minimum requirements for the discharges of waste water into waters. It is thus ensured that the discharge of hazardous substances from areas of industry without formulated minimum requirements for guaranteeing the proper operation of a sewage treatment plant can be limited by the local municipal statutes. For some areas of industry the state of the art with regard to residual concentrations of hazardous substances which can be achieved is not yet known or laid down. The result of this is that in A 115 the discharge concentrations have been set at about twice the value given in the appendices to the general administrative framework regulation on minimum requirements for the discharges of waste water into waters.

Table 45: Recommended values for hazardous substances according to leaflet A 115 compared to the maximum values of the minimum requirements of the appendices to the general administrative framework regulation on minimum requirements for the discharge of waste water into waters (mg/l)

Parameter	Recommended value ATV A 115	Minimum requirement of the general administrative framework regulation
Ag	1	0.7
As	0.5	0.3
Ba	5	3
Cd	0.5	0.2
Cp	2	1
Cr total	1	1
Cr(VI)	0.2	0.1
Cu	1	0.5
Hg	0.1	0.05
Ni	1	0.5
Pb	1	0.5
Sb	0.5	0.3
Se	2	1
Sn	5	2
Zn	5	2
CN, easily released	1	1
Sulfide S	2	1
AOX (adsorbable organic halogen)	1	1
Volatile halogenated hydrocarbons	0.5	0.1

7.7   Soil

The testing of soil can provide important information on nutrient deficiency or excess. For tropical and subtropical soils the specialist literature must be referred to (e.g. Mohr, Landon).

For soils with a high salt content recommendations concerning fertiliser treatment can only be made after the whole mineral content has been determined. Cultivated crops in salty soil suffer not only from a lack of water, but often also from nutrient problems. Alkaline salty soils have an SAR value of more than 13 (see Section 6.3.2.8) and their aqueous extracts (1:5) an electrical conductivity of more than 4 mS cm^{-1}. For soils rich in alkaline earths the corresponding values are SAR < 13 and electrical conductivity > 4 mS cm^{-1}.

Heavy metals are frequently introduced into soil via domestic sewage sludge. The recommended and threshold values listed in Table 46 for the spreading of sewage sludge are in force in Germany and the rest of the EC.

The testing of sludge according to the Sewage Sludge Ordinance requires the testing for heavy metals and for the percentage concentrations of the following nutrients in both the fresh and dry substance if the sludge is intended for agricultural use:

- organic materials,
- total nitrogen,
- ammonium nitrogen,
- phosphate (as P_2O_5),
- potassium (as K_2O),
- calcium (as CaO),
- magnesium (as MgO).

In soils the following are also examined: phosphate, potassium and magnesium, given in mg/100 g dry substance. The German Fertiliser Ordinance of 1996, which replaces the EEC Guideline 91/676/EWG of 1991 for the protection of bodies of water from contamination by nitrate from agricultural sources, may be referred to.

The laying of pipes in the ground requires that certain soil characteristics, which can affect the corrosion of iron and steel, are considered. Sometimes protective measures need to be taken during laying. The most important parameters to be considered are as follows.

Soil Texture
Sandy and lime-containing soils and well aired loams are generally not aggressive. Peaty, lime-free humus soils and silted soils are aggressive, as are deposited soils (slag, refuse).

Soil Humidity
In aggressive soil the corrosion is greatest at a water content of ca. 20%.

pH
In soils with a pH of < 6 (measured as a suspension in distilled water) aggressiveness increases with decreasing pH.

Total Acidity

Soils with pH < 7, where the total acidity is greater than the consumption of 25 ml sodium hydroxide (c (NaOH) = 0.1 mol/l) per kg soil, are classified as aggressive.

Lime Content

Aerobic soil with a calcium carbonate content of more than 5% is not aggressive at low sulfate concentrations.

Carbon

Soils containing elemental carbon are considered to be aggressive because of the danger of formation of electrochemical cells.

Table 46: Threshold and recommended values for the spreading of sewage sludge on agricultural land, mg/kg dry substance

Substance	Sewage Sludge Ordinance of 1992	EEC Guideline 86/278/EWG
a)[1]		
Cd	10 (5)*	20–40
Cr	900	–
Cu	800	1000–1750
Hg	8	16–25
Ni	200	300–400
Pb	900	750–1200
Zn	2500 (2500)*	2500–4000
PCB	0.2	
AOX (adsorbable organic halogen)	500	
Dioxins/furans	100 ng TE/kg	
b)[2]		
Cd	1.5 (1)*	
Cr	100	
Cu	60	
Hg	1	
Ni	50	
Pb	100	
Zn	200 (150)*	

[1] Sewage sludge may only be spread on agricultural land without official permission if the concentrations of the heavy metals listed are not exceeded.

[2] The spreading of sewage sludge on agricultural land is forbidden if at least one of the values given is exceeded in the soil tested.

* Stricter threshold values for light soils and those with pH 5–6.

Chloride

Chloride concentrations (measured in an aqueous extract) of more than 100 mg/kg promote corrosion.

Sulfate

Sulfates can promote corrosion if their content (measured in an aqueous extract) exceeds 200 mg/kg. In aerobic soils with calcium carbonate contents of more than 5%, sulfate concentrations of up to 500 mg/kg are harmless.

8   References

Books

American Public Health Association (ed.) (1989): Standard Methods for the Examination of Water and Wastewater. APHA, Washington D.C.

American Water Works Association (ed.)(1997): Water Treatment Plant Design. McGraw Hill, New York.

ASTM (ed.) (1991): Annual Book of ASTM Standards, Section 11: Water and Environmental Technology. ASTM, Philadelphia.

Atlas R. M. (1996): Handbook of Media for Environmental Microbiology. CRC Press, Boca Raton (USA).

Blume H. P. (1992): Handbuch des Bodenschutzes. Ecomed, Landsberg. (Handbook of soil protection).

Bock R. (1979): Handbook of Decomposition Methods in Analytical Chemistry. International Textbook Company, London.

Cheeseman R., A. Wilson (1978): Manual on Analytical Quality-Control for the Water Industry. Water Research Centre, Stevenage (UK).

Environmental Protection Agency (ed.) (1979): Methods for Chemical Analysis of Water and Wastes. EPA, Cincinnati.

Fachgruppe Wasserchemie in der GDCh (Hrsg.): Deutsche Einheitsverfahren zur Wasser-, Abwasser- und Schlammuntersuchung. Ergänzungswerk Lose-Blatt-Sammlung, Verlag Chemie, Weinheim. (Specialist group for water chemistry in the German Chemical Society (publisher): German standards for the examination of water, waste water and sludge, supplement to the loose leaf collection, Verlag Chemie, Weinheim).

FAO/UNESCO (ed.) (1973): Irrigation, Drainage and Salinity. FAO, Paris.

Förstner U., G. Wittmann (1981): Metal Pollution in the Aquatic Environment. Springer, Berlin.

Fresenius W., K. E. Quentin, W. Schneider (eds.)(1988): Water Analysis. Springer, Berlin.

Frimmel F., R. Christman (eds.) (1988): Humic Substances and their Role in the Environment. Wiley, New York.

German Federal Ministry for Economic Cooperation and Development – BMZ (ed.)(1995): Environmental Handbook. Vol. 1–3, Vieweg, Braunschweig.

Hutton L. (1983): Field Testing of Water in Developing Countries. Water Research Centre, Medmenham.

Institut Fresenius/FIW TH Aachen (Hrsg.) (1989): Wastewater Technology. Springer, Berlin.

Landon, J. R. (1984): Booker Tropical Soil Manual. Longman, Harlow.

Masschelein W (ed.) (1982): Ozonation Manual for Water and Waste Water Treatment. Wiley, New York.

Mohr E. C. et al. (1992): Tropical Soils. A Comprehensive Study on their Genesis. Mouton, The Hague.

Oliveira M. (1996): Praxishandbuch Laborleiter. WEKA, Augsburg. (Manual for laboratory managers).

Petermann T. (1993): Irrigation and the Environment. Parts I and II., GTZ, Eschborn.

Richards L. (ed.) (1969): Diagnosis and Improvement of Saline and Alkali Soils. US Dep. of Agriculture, Washington.

Riikonen N., C. Jones (1992): Industrial Wastewater Source Control – An Inspection Guide. Technomic Publishing, Lancaster Pa.

Römpp Chemie-Lexikon (1989-1992): Thieme, Stuttgart.

Rump H. H., B. Scholz (1995): Untersuchung von Abfällen, Reststoffen und Altlasten. VCH, Weinheim. (Examination of wastes, residues and hazardous waste sites).

Sanchez P. (1976): Properties and Management of Soils in the Tropics. Wiley, New York.

Scheffer F., P. Schachtschabel (1992): Lehrbuch der Bodenkunde. Enke, Stuttgart. (Handbook of soil Science).

Schnitzer, M., S. Khan (1978): Soil Organic Matter. Elsevier, Amsterdam.

Schulz S., D. Okun (1984): Surface Water Treatment for Communities in Developing Countries. Wiley, New York.

Smith, K. (1991): Soil Analysis. Dekker, New York.

Spillmann P., H.-J. Collins, G. Matthess, W. Schneider (Hrsg.) (1995): Schadstoffe im Grundwasser. DFG-Forschungsbericht Bd. 2: Langzeitverhalten von Umweltchemikalien und Mikroorganismen aus Abfalldeponien im Grundwasser. VCH, Weinheim. (Pollutants in groundwater, part 2: long-term behaviour of environmental pollutants and microorganisms from landfills in groundwater).

Suess M. (WHO) (ed.) (1982): Examination of Water for Pollution Control. Vol. 1–3. Pergamon Press, Oxford.

Stumm W., J. Morgan (1996): Aquatic Chemistry. 3 rd Ed. Wiley, New York.

Townshend A. (ed.) (1995): Encyclopedia of Analytical Science. Academic Press, London.

Ullmann's Encyclopedia of Industrial Chemistry (1985–1997): 5. Edition, Vol. B5+B6. VCH, Weinheim.

Weast R. (ed.) (1990): Handbook fo Chemistry and Physics. The Chemical Rubber Co., Cleveland.

World Health Organisation (1989): Disinfection of Rural and Small Community Water Supplies. WHO, Copenhagen.

Laws, Ordinances, Guidelines

Gesetz zur Ordnung des Wasserhaushalts (Wasserhaushaltsgesetz – WHG) vom 27.07.1957, Bundesgesetzblatt Teil I, S. 1110; zuletzt geändert am 27.06.1994, Bundesgesetzblatt Teil I, S. 1440. (Federal Water Act).

Gesetz zum Schutz vor schädlichen Umwelteinwirkungen durch Luftverunreinigungen, Geräusche, Erschütterungen und ähnliche Vorgänge, Bundes-Immissionsschutzgesetz – BImSchG, Neufassung vom 14.05.1990, Bundesgesetzblatt Teil I, S. 880; zuletzt geändert am 26.09.1994, Bundesgesetzblatt Teil I, S. 2640. (Law for the protection against damage to the environment by air pollution, noise, vibrations and similar processes (Federal Immission Control Act)).

Gesetz zum Schutz vor gefährlichen Stoffen (Chemikaliengesetz – ChemG), vom 16.09.1980, Bundesgesetzblatt Teil I, S. 1718; zuletzt geändert am 29.07.1994, Bundesgesetzblatt Teil I, S. 2705. (Law for the protection against harzardous substances (Chemicals Act)).

Gesetz zur Förderung der Kreislaufwirtschaft und Sicherung der umweltverträglichen Beseitigung von Abfällen (Kreislaufwirtschafts- und Abfallgesetz – KrW/AbfG) vom 27.09.1994, Bundesgesetzblatt Teil I. S. 2705. (Law for the promotion of recycling and ensuring the environmentally friendly disposal of waste (Closed Substance Cycle and Waste Management Act)).

Gesetz über Abgaben für das Einleiten von Abwasser in Gewässer (Abwasserabgabengesetz – AbwAG), vom 03.11.1994, Bundesgesetzblatt Teil I, S. 3370. (Law governing the discharge of waste water into bodies of water (Waste Water Charges Act)).

Klärschlammverordnung (AbfKlärV) vom 15.04.1992, Bundesgesetzblatt Teil I, S. 912. (Sewage Sludge Ordinance).

Verordnung über Trinkwasser und über Wasser für Lebensmittelbetriebe (Trinkwasserverordnung – TrinkwV) vom 05.12.1990, Bundesgesetzblatt Teil I, S. 2612, zuletzt geändert am 26.2.1993, Bundesgesetzblatt Teil I, S. 227. (Ordinance on drinking water and water for use in foodstuffs).

Verordnung zum Schutz vor gefährlichen Stoffen (Gefahrstoffverordnung – GefStoffV) vom 26.10.1993, Bundesgesetzblatt Teil I, S. 1782; zuletzt geändert am 19.09.1994, Bundesgesetzblatt Teil I, S. 2557. (Ordinance for protection against hazardous substances (Hazardous Materials Ordinance)).

Zweite Allgemeine Verwaltungsvorschrift zum Abfallgesetz, Teil 1: Technische Anleitung zur Lagerung, chemisch/physikalischen, biologischen Behandlung, Verbrennung und Ablagerung von besonders überwachungsbedürftigen Abfällen (TA Abfall) vom 12.03.1991, Gemeinsames Ministerialblatt, S. 139. (Second general government regulation of the Closed Substance Cycle and Waste Management Act, part 1: Technical instruction on the storage, chemical/physical, and biological treatment, incineration and disposal of waste requiring special monitoring (Technical Instructions On Waste)).

Erste Allgemeine Verwaltungsvorschrift zum Bundes-Immissionschutzgesetz, Technische Anleitung zur Reinhaltung der Luft – TA Luft vom 27.02.1990, Gemeinsames Ministerialblatt S. 95; zuletzt geändert am 04.04.1996, Gemeinsames Ministerialblatt, S. 202. (First general government regulation of the Federal Immission Control Act; Technical instruction for the maintenance of air purity).

EG-Richtlinie 80/68/EWG: Richtlinie des Rates über den Schutz des Grundwassers gegen Verschmutzung durch bestimmte gefährliche Stoffe, Abl. L 20/43. (EC guideline 80/68/EEC: guideline of the advisory body concerning the protection of ground water against contamination by particularly hazardous substances).

EG-Richtlinie 75/440/EWG: Richtlinie des Rates vom 16.06.1975 über die Qualitätsanforderungen an Oberflächengewässer für die Trinkwassergewinnung in den Mitgliedsstaaten, Abl. L 194 v. 25.07.75, S. 34. (EC guideline 75/440/EEC: Guideline of the advisory body of 16.06.1975 concerning the quality requirements for surface water for use in drinking water preparation in the member states).

EG-Richtlinie 98/83/EG: Richtlinie des Rates vom 3.11.1998 über die Qualität von Wasser für den menschlichen Gebrauch. Abl. L 330 v. 5.12.1998, S. 32. (Council directive 98/83/EC of 3 Nov. 1998 on the quality of water intended for human consumption).

EWPCA Task Group (1990): Wastewater Quality Standards in European Countries. Water Quality Institute, Denmark.

WHO (1993): Guidelines for Drinking Water Quality. Vol. 1: Recommendations. WHO, Geneva.

TRGS 102 (1993): Technische Richtkonzentrationen (TRK) für gefährliche Arbeitsstoffe, Bundesarbeitsblatt Nr. 1/1994. (Technical Guide Concentrations (TRK) for hazardous materials).

TRGS 402 (1986): Ermittlung und Beurteilung der Konzentrationen gefährlicher Stoffe in der Luft in Arbeitsbereichen, Bundesarbeitsblatt Nr. 11/1986 und Nr. 10/1988. (Determination and assessment of the concentrations of hazardous substances in the air at the workplace, official leaflet).

TRGS 403 (1989): Bewertung von Stoffgemischen in der Luft am Arbeitsplatz, Bundesarbeitsblatt Nr. 19/1989. (Assessment of substance mixtures in the air at the workplace).

TRGS 555 (1989): Betriebsanweisung und Unterweisung nach §20 GefStoffV, Bundesarbeitsblatt Nr. 3/1989 und Nr. 10/1989. (Operating instructions according to § 20 of the Hazardous Materials Ordinance).

TRGS 900 (1994): Grenzwerte in der Luft am Arbeitsplatz – MAK- und TRK-Werte, Bundesarbeitsblatt Nr. 6/1994. (Threshold values in the air at the workplace – MAK and TRK values).

TRGS 905 (1994): Verzeichnis krebserzeugender, erbgutverändernder oder fortpflanzungs-gefährdender Stoffe, Bundesarbeitsblatt Nr. 6/1994. (List of substances which are carcinogenic, or harmful to genetic material or to reproduction).

VBG 100: UVV Arbeitsmedizinische Vorsorge. (Industrial medical precautions).

DVGW-Merkblatt W 121: Bau und Betrieb von Grundwasserbeschaffenheitsmessstellen. (The building and operation of measuring points for assessing ground water).

DVGW-Merkblatt W 112: Entnahme von Wasserproben bei der Wassererschliessung. (Collection of samples from water sources).

DVWK-Merkblatt 208: Entnahme von Proben für hydrogeologische Grundwasseruntersuchungen. (Collection of samples for hydrogeological ground water investigations).

9   Index